A BIRDWATCHERS'
GUIDE TO THE GAMBIA

ROD WARD

Illustrations by Rob Hume

BIRD
WATCHERS'
GUIDES

Prion Ltd.
Perry

ACKNOWLEDGEMENTS

I am indebted to so many people, both in the UK and in The Gambia, for assistance, information and advice, that it is impossible to mention all of them by name. In particular I should like to thank Dr Torben Larsen for kindly sending me information on butterflies of The Gambia, and Dr E.N.Arnold and Dr B.T.Clarke of the Natural History Museum for help with the reptiles and amphibians respectively. Dr Barry Blake and Dr Tony Walker provided assistance in many ways, as did David Hennessy, and I am grateful to all of these. I should like to thank all my Gambian friends for making me so welcome in their country, and all my customers whilst I was there for giving me the opportunity to visit so many sites so many times. I am especially grateful to Phil Smith, without whose spirit of adventure I might never have discovered the delights of the Faraba Banta - Jiboroh Kuta bush track. Finally, and most of all, I should like to thank my wife, Lynn, for her tolerance, understanding and help in so many ways.

Rod Ward has been birdwatching for 20 years and has travelled widely in Europe, Africa, the Middle East and Central America. He has visited The Gambia on several occasions in winter and spring both birdwatching and guiding birdwatchers around. He has travelled throughout The Gambia and has a wide knowledge of the country and the people from a tourist's and a tour operator's point of view.

CONTENTS

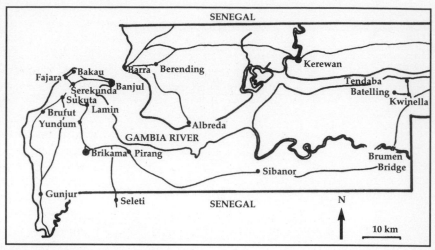

The Gambia – Lower River

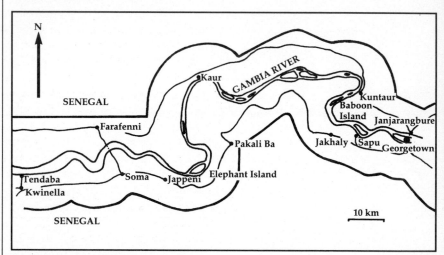

The Gambia – Middle River

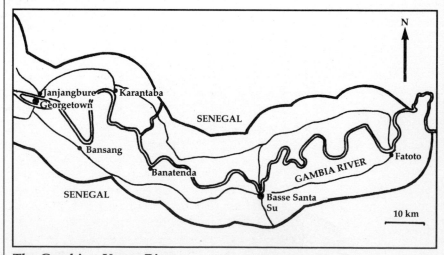

The Gambia – Upper River

INTRODUCTION

The Gambia comprises a long, narrow strip of land, about 50km across at its widest point and just over 300km long, bordered to the west by the Atlantic Ocean and otherwise completely surrounded by Senegal.

The Gambia River, which divides the country into two, is tidal inland for about 200km and this explains the abundance of mangroves, which cover almost 30 per cent of the total land area. White Mangroves grow above the normal high tide level where pneumatophores (breathing roots) can be seen sticking up above the mud. Larger Red Mangroves are found in wetter areas and have stilt roots but no pneumatophores.

North of the river, desertification has led to the replacement of much of the savannah by desert scrub, and that remaining is more open Sudan savannah rather than the denser, wooded Guinea savannah once typical of the Coastal and Lower River regions on both sides of the river. Some Guinea savannah remains on the south bank, but much of the area not covered by mangroves has been cultivated, resulting in what is often called 'parkland savannah' - fairly open countryside with scattered large trees such as silk-cottons, baobabs and acacias. Little original gallery forest (also called riverine forest) remains in the country, but there is a readily accessible example at the Abuko Nature Reserve. Add to this variety of habitats the coastal region and the flower-filled gardens of the tourist areas and it is no wonder that the country has such a rich avifauna.

The Gambia was a British colony until independence in 1965, becoming The Republic of The Gambia in April 1970. The President since independence has been Sir Alhaji Dawda Kairaba Jawara, leader of the People's Progressive Party (PPP), the largest party in the governing National Assembly. There is complete political freedom and freedom of speech in The Gambia, which is very proud of its democratic record. The official language of The Gambia is English, which is spoken by virtually everyone in the tourist areas although older people, especially up-country, may speak only their tribal language.

Although the majority of the population is Moslem, the Gambian form of Islam is much less overt than that found in the Middle East or even in the countries of North Africa, and for most of the time you will see few outward signs of the dominant religion. Women do not wear veils, alcohol is freely available in tourist areas, and the occasional figure in traditional Moslem dress is the exception rather than the rule.

Serious crime is rare and, although there are some places where it is not advisable to venture at night, in general you are as safe walking around as you would be in the streets of your own home town. However, as in many affluent tourist destinations, petty theft, pick-pocketing and handbag-snatching are a problem wherever there are crowds and it pays to take precautions with valuables.

The country is small, making comprehensive coverage of the main birding sites a real possibility, even in two weeks. Communications are adequate, if sometimes erratic, and the chances of being completely stranded are remote. Add to these advantages a list of over five hundred reliably recorded bird species, and a good chance of seeing at least half that number of them in a two-week visit, and it is no wonder that the country is one of the most popular birding destinations in the world.

PRE-TOUR INFORMATION

Visas and passports

Citizens of the United Kingdom and the British Commonwealth do not require a visa to visit The Gambia, but a full ten year passport is essential. Nationals of other countries may be subject to different entry requirements and should contact their nearest Gambian Embassy, Consulate or High Commission for information. The address to write to in Britain is: The Gambian High Commission, 57 Kensington Court, London W8 5DG.

At the time of writing, holders of Israeli or South African passports are refused entry, but recent events in both countries may well lead to changes in the near future.

When passing through immigration, passports will be stamped allowing a length of stay equivalent to that stated on the immigration form filled in on the plane. If you have requested more than the standard two weeks, check that it has been stamped correctly - it is easier to have it corrected on the spot rather than leaving it until later. For stays of longer than one month a visit may be needed to the aliens section on the ground floor of the Immigration Offices, on the corner of Dobson Street and Anglesea Street in Banjul. A form must be filled in giving the reason for wanting an extension ('holiday' is quite sufficient). Anyone still in the country at the end of three months, will be required to write a letter to the Minister of Immigration in person, explaining the purpose of the visit and the reason for staying longer.

Currency and Exchange Rate

The unit of currency is the dalasi, divided into 100 bututs. Banknotes, usually filthy and in an advanced state of disintegration, are for 50, 25, 10 and 5 dalasis, while coins are 1 dalasi and 50, 25, 10 and 5 bututs.

The exchange rate fell steadily through the 1993 season and was about D12 in March 1994. The rate given by hotels varies from a fraction below the bank rate at, for example, the Bakotu and The Bungalow Beach, to a very poor rate at most of the larger hotels. The rate at the official exchange offices is better than that given by the hotels and about a point below the banks, but even the different banks themselves do not seem able to agree on a uniform rate.

Street money-changers near the Albert Market in Banjul, and sometimes outside the hotels at Kotu, are perfectly legitimate, but only give their best rates for relatively large sums of money which it may not be wise to be seen changing in public.

On no account buy dalasis in the UK before departure, as the rate offered is laughable. You are unlikely to need any Gambian currency before you reach your hotel, where the rate, however poor, will be infinitely better than anything from a bank at home. If you need to tip a porter, give a British coin. These are quite acceptable since departing tourists happily accept them in exchange for dalasis or bututs. If you are travelling independently, you will need to pay for a taxi from the airport but even so you are unlikely to need more than about D200. In emergency, sterling bank notes can be used as they are accepted readily, but check the rate at the airport bank.

There is no restriction now on the import or export of currency, Gambian or otherwise, but it is as well to either spend or exchange remaining dalasis before leaving the country as they are worthless outside The Gambia.

Field Guides

'A Field Guide to the Birds of West Africa' by Serle, Morel and Hartwig is at present the only comprehensive bird guide for the area but leaves a lot to be desired, with many illustrations in inappropriate black and white. Newman's 'Birds of Southern Africa' is a useful addition, being well illustrated in full colour and including many species from the Gambia list, but is unfortunately rather bulky. 'A Field Guide to the Birds of East Africa' by Williams and Arlot also contains several relevant colour illustrations. 'Birds of Europe, with North Africa and the Middle East' by Lars Jonsson is useful for Palaearctic visitors and passage migrants and also covers several relevant African species.

Photography

The Gambia presents many opportunities for photography and no major problems will be encountered provided the visitor bears the following in mind:

1. Take plenty of film. Film is available but is not cheap and not all brands are easily found.
2. When it is not in use, carry photographic equipment in a plastic bag. Not only are you less obtrusive but, more importantly, your gear will be protected against the all-pervading red laterite dust.
3. Don't photograph people without their permission. Children usually love it but older people may object strongly, not because they think you are stealing part of their soul but simply because they suspect you may be going to sell the result and thus make money at their expense. A refusal to be photographed must be respected or an unpleasant confrontation may develop - the offer of a dalasi or so may bring about a change of mind.
4. Don't take photographs at police stations, border posts, army depots or at the airport. Your film will probably be confiscated and possibly your camera as well. In such places it is best to keep cameras in their cases, or well out of sight, to avoid any possible misunderstandings.

Health certificates and inoculations

The Gambia is a tropical country with all the attendant potential and health hazards. There are no compulsory inoculations for visitors but prophylaxis against yellow fever, hepatitis 'A', typhoid, tetanus, malaria and polio is strongly recommended. Inoculation against cholera seems no longer to be recommended, but check with your doctor.

Emergencies

In case of emergency, contact: The UK High Commission, 48 Atlantic Road, Fajara. Postal address: PO Box 507, Banjul, The Gambia. Telephone: 95133, 95134 or 95578.

TRAVEL INFORMATION

Travelling in The Gambia

Most visitors arrive on standard package deals, which take care of both accommodation and travel to and from The Gambia while leaving you free to organize your birdwatching as you wish. With the exception of Basse and Tendaba, all the sites described can be visited in a day from any of the tourist hotels, which are all located within about half an hour's drive of one another.

Several UK operators offer holidays to The Gambia, most during the winter season only but some all year round. The Gambia Experience are specialists in holidays to the country and offer some very good 'out-of-season' deals. Their brochure is not usually available through travel agents but can be obtained by writing direct to: The Gambia Experience, 28 The Hundred, Romsey Hampshire, SO51 8BW, or telephoning (0794) 830999. The company also have an office in The Gambia, run by extremely helpful staff and located in the Kairaba Hotel (Tel. 90317).

Independent travel is not really recommended unless you already know The Gambia well since, apart from the tourist hotels, European-style accommodation is in short supply and not easy to locate. Outside the peak tourist period, which occurs around Christmas and New Year, there should be few problems finding a hotel room but, added to the cost of the flight and transport to and from the airport, this will almost certainly work out more expensive than a complete package.

Following the demise of Air Gambia in 1994 there are no direct scheduled flights from the UK to The Gambia. SwissAir currently operate flights from London, Birmingham and Manchester, via Zurich, on Fridays while the Belgian airline, Sabena, fly from both London and Manchester, via Brussels, on Mondays, Wednesdays and Fridays.

A cheaper alternative to a scheduled flight is a flight-only deal with a charter airline. These are at present available through Thomson Holidays, Skybargains and Falcon Flights during the winter season, and through The Gambia Experience throughout the year. The latter, however, have a policy of only providing this service to people who are already well-acquainted with the country.

International Airport

The only airport in The Gambia is Banjul International Airport, situated at Yundum, some 25 km or so from Banjul itself. Branches of The Gambia Commercial and Development Bank are located at arrivals and departures and there is a small duty free shop. There are no car hire facilities at present.

There is no public transport to or from the airport, and the nearest buses are a long walk away down the airport road, so if you are on a flight only deal you will have to get a taxi (D180 one way) or beg a lift. Don't wait around too long as it is not a busy airport and is fairly deserted between flights.

Travelling within The Gambia

Taxis

Tourist taxis can be found outside all the main hotels. Prices are fixed and the operations strictly controlled, although it may sometimes be possible to strike a private deal with a driver away from the depot. They are an expensive way of getting around if you are on your own or with just one companion but for a small group sharing the cost they are a useful way of reaching less accessible sites such as the Tanji Reserve. A full day out locally should cost about D500 (for the vehicle, not per person), but trips up-river are considerably more expensive. One of the most helpful and reliable drivers in the Kotu area is Mr Roberts (he appears to have no other name). He runs a particularly smart and comfortable Peugeot and can be booked in advance. His telephone numbers are 96691 (10.00 to 19.00) and 91873 (20.00 to 07.00). When free, he is usually on the carpark next to the Bakotu Hotel during the day.

Many of the best sites really need a four-wheel drive vehicle, especially the bush track from Faraba Banta to Jiboroh Kuta and the woods at Kabafita and Brufut. Landrovers with driver/guide can be chartered from a number of companies, including Crocodile Safaris (Tel. 96068) and Black & White Enterprises (Tel. 93174), both of which are long-established and reliable concerns and can easily be contacted through your hotel reception. A number of independent operators run open-topped Suzuki jeeps, which are ideal for birdwatching since they can be used as 'mobile hides'. Several such operators can be found around the Kotu Beach area and ones who can be trusted to bring you back in one piece include 'Big Mike' (Tel. 96111) and Morfa Sunneh (Tel. 94326).

Car Hire

Self-drive is not recommended unless you are used to driving in West Africa. Outside the tourist areas the roads are often of very poor quality and the standard of driving even worse. If you remain undeterred, then Avis have an agency on Atlantic Avenue, Bakau (Tel. 96119) and there is a Hertz agency at the Senegambia Hotel. You must be over twenty five, and a full British driving licence is valid for up to three months.

Buses

Buses are operated by GPTC (Gambia Public Transport Corporation) and run on a limited number of routes, linking Banjul with Soma, Basse, Serekunda, Bakau, Tanji and Gunjur and running on the north bank from Barra to the Senegalese frontier and to Georgetown via Kerewan and Farafenni. There are bus stops at intervals along the routes but buses are often full. The best way to make sure of a seat on an outward journey is to board at the Banjul terminus which is located in Cotton Street, at the end of the Bund Road (shown on the map of Banjul p10).

'Bush Taxis'

Once you have mastered the art of catching one, you will find that bush taxis, known locally as 'tankatanks', are by far the cheapest and most convenient way of getting around. Like tourist taxis they have yellow number plates, but there all similarity ends. Bush taxis operate on fixed routes and frequently lack such inessential luxuries as driving

Travel Information

mirrors, door-handles, shock-absorbers and so on. Cars operate on the shorter routes and minibuses or converted pickups on the longer or more popular ones. They cost only a few dalasis and pick people up anywhere along their journey. Some bush taxis are reluctant to stop for obvious tourists so it is important to try to look as little like one as possible. Avoid shorts and beach shirts, and keep cameras and binoculars out of sight in a plastic carrier bag. Don't try to board or stop a bush taxi directly outside an hotel - tourist taxi drivers object strongly if others appear to be pinching their trade. Instead, walk a little way along the road, watch out for a yellow number plate approaching and confidently flag it down. If the vehicle is empty, check that it is not a tourist taxi or you may be left with a larger bill than anticipated. Be careful, too, if the driver asks you where you want to go; if you mention somewhere off the normal route he may take you there, but you will be charged extra. It is a good idea to familiarize yourself with the main bush taxi routes in advance so that you know exactly where you are going; those into Serekunda from the tourist areas are shown below. Serekunda is a good starting place for trips further afield, since bush taxis converge here from virtually everywhere and leave for destinations far and wide. The map of Serekunda (p10) shows departure points for various destinations while the main drop off/pick up point in Banjul is shown on the city map (p10).

Bicycles

Mopeds and bicycles can be hired at Kotu and elsewhere. Charges vary depending on the whim of the owner, but they are not expensive.

Hitching

Hitch hiking is not usual, and if you are picked up you will probably be expected to pay bush taxi rates.

Key

Bush taxi routes:
Serekunda – Kololi
Serekunda – Kotu
Serekunda – Fajara
– Bakau

Hotels
Holiday Beach Club	A
Kairaba	B
Senegambia Beach	C
Palma Rima	D
Badala Park	E
Kotu Strands	F
Kombo Beach	G
Bungalow Beach	H
Bakotu	I
Fajara	J
Francisco's	K
African Village	L
Amie's Beach and Sunwing	M
To Wadner Beach, Palm Grove and Atlantic	N

Bush taxi routes to Serekunda

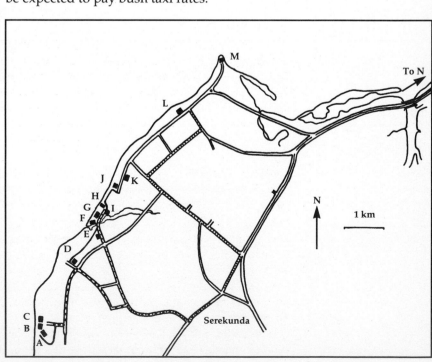

Accommodation

Hotels   Tourist hotels range from the large and luxurious, such as the Senegambia Beach, to the small, comfortable and friendly, such as the Bakotu, with a range of prices to match. The hotels are spread along the coast from Banjul to Kololi and some are more conveniently situated than others as far as birdwatchers are concerned. The following brief notes may assist in the selection of a suitable base.

Kololi Area

Hotels in this area are a short walk from Bijilo Forest Park and about fifteen minutes along the beach from the Palma Rima end of the Casino Cycle Track, along which the Kotu area can be reached on foot. Bush taxis leave for Serekunda from the Kololi depot just up the road, near the Gamtel office (from which local and international telephone calls can be made) and a branch of the Standard & Chartered Bank.

Kairaba (Tel. 92940): The most luxurious and expensive hotel in The Gambia.

Senegambia Beach (Tel. 92718): A big, busy, luxury hotel with extensive grounds which offer rewarding birdwatching without leaving the premises.

Holiday Beach Club (Tel. 90418): Pleasantly situated alongside the beach close to Bijilo Forest Park and the Kololi Beach Club.

Kotu Beach

Close to Kotu Creek, Kotu Ponds, the Fajara Golf Course and the Casino Cycle Track this is probably the best base for a birdwatching holiday. The main Kotu Beach area has a Gamtel office, a currency exchange, an African Craft Market and several mini-supermarkets, as well as a number of beach bars and restaurants and a bush taxi depot. It can be noisy at night during the main season, but the excellent birding sites within easy walking distance more than compensate for this.

Palma Rima (Tel. 93380): A large hotel at the southern end of the Casino Cycle Track just over 1km from Kotu Creek and about 2km along the beach from Bijilo Forest Park. The extensive area of coastal scrub between the hotel and the shore is excellent for birds. Bush taxis pass the hotel en route between Kololi and Serekunda.

Badala Park (Tel. 90400): On the road to the main Kotu Beach area, a short distance from Kotu Creek and alongside the Casino Cycle Track. Next door to the 'Tam-Tam' nightclub and almost opposite the Kotu Ponds sewage treatment plant. Bush taxis running between Kotu and Serekunda pass by fairly regularly.

Kombo Beach (Tel. 95466): A large and popular Novotel hotel in the main Kotu Beach area with an excellent a la carte restaurant, although drinks from the bar are the most expensive in the area. Noisy evening entertainment in the high season.

Bakotu (Tel. 95555): A small, friendly hotel opposite the African Market, set in beautiful gardens which attract a wide variety of birds. There is a birdwatching hide in the grounds and a small, shady terrace overlooking parts of the Fajara Golf Course and Kotu Creek. The one drawback to this relatively inexpensive hotel, in an excellent birdwatching area, is the drums of the Kotu Bendula Bar in the African Market opposite, which go on late during the high season. Rooms in G Block, or one of the others at the far end of the gardens, are the quietest.

Bungalow Beach Hotel (Tel. 95288): Known to all as the 'BB', this hotel consists of self-catering apartments set in grounds which run down to the beach. The staff are friendly and helpful and there is a good restaurant and a pool-side bar. It suffers from the over-enthusiastic drummers at the Kotu Bendula but rooms 27 or 28 in any of the blocks (A, B or C) are at the beach end of the grounds and have the added advantage of sea watching from the upper level balcony and a pleasant breeze off the sea for much of the time.

Fajara

Fajara is a smart residential area at the southern end of Atlantic Road, close to the Golf Course and within easy reach of Kotu Creek and the surrounding areas as well as the MRC grounds and the rest of Atlantic Road (sometimes, confusingly, also called Atlantic Avenue).

Aparthotel Fajara (Tel. 95339): Newly renovated chalet-type accommodation set in grounds which slope down to the Kotu beach.

Francisco's (Tel. 95258): A small, friendly guest house and restaurant, run by a British couple and close to the Fajara Club and Golf Course. The eight bedrooms have fans and refrigerators at no extra charge.

Bakau

Bakau lies about 2km north-east of Fajara along Atlantic Road.

African Village (Tel. 95384): A Gambian-run hotel set in pleasant gardens alongside the shore at Bakau. It is close to the native market, the CFAO supermarket, a Gamtel office and two banks, with the bush taxi depot almost opposite. The Old Cape Road starts about 1km up the road and the MRC grounds are about the same distance in the opposite direction.

Cape Point

Cape Point is close to the Old Cape Road but is otherwise not very well-situated for birdwatchers who wish to explore independently.

Amie's Beach Hotel and Apartments (Tel. 95035): A large hotel on Cape Point and with several bars and restaurants.

Sunwing (Tel. 95428): Set amongst attractive gardens and close to the Amie's Beach on Cape Point.

Banjul Island (St Mary's Island)

Wadner Beach (Tel. 28199): Situated alongside the main Serekunda-

Banjul highway about 3km from the centre of Banjul. It is about 1km from the northern end of the classic Bund Road birding area.

Palm Grove (Tel. 28630): Next door to the Wadner Beach and also rather isolated and self-contained.

Atlantic (Tel. 28601): Located on Marine Parade about a mile from the centre of Banjul, this large, top class hotel is set in extensive gardens with a wide range of facilities. It is not in an ideal location for tourists but the Gambian ornithologist Mass Cham can be contacted through the hotel reception, and the northern end of the Bund Road is only a short walk away. Reaching other birdwatching destinations will require either a tourist taxi from outside the hotel or a walk into Banjul to pick up a minibus bush taxi. These pass the Old Cape Road and Camaloo Corner on the way to Serekunda, from where other taxis leave for Kotu, Kololi, Abuko and elsewhere.

Apartments

Privately rented European-style rooms and apartments are available from as little as D140 per night for the accommodation (not per person) but quality and facilities vary considerably. Since advertisement is usually by word of mouth it is best to have a contact in the country to make arrangements. At Banjul airport there may be people around the terminal building offering rooms or apartments and these are mostly quite genuine offers and may prove to be good deals, but always inspect the accommodation before becoming committed.

Camping

There are no camp sites in The Gambia, and camping in the open is definitely not recommended.

Egyptian Plover

Key

Standard Chartered Bank	A
Supermarket	B
Minibus Bush Taxis for Abuko, Yundum, Brikama	C
Bush Taxis for Fajara and Bakau	D
Mosque	E
Shell Garage	F
Bush Taxis for Kotu and Kololi	G
Minibus Bush Taxis for Banjul	H

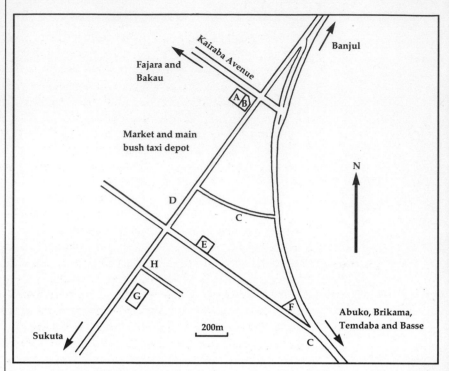

Central Serekunda

Key

Atlantic Hotel	A
Great Mosque	B
Royal Victoria Hospital	C
National Museum	D
Anglican Cathedral	E
Government Offices (Quadrangle)	F
War Memorial	G
Albert Market	H
Gamtel Office	I
GPO	J
Minibus departure point (for Serekunda)	K
Minibus arrival point (from Serekunda)	L
Immigration Offices	M
CFAO Supermarket	N
African Heritage Restaurant	O
Ferry Terminal (Barra)	P
Bus Terminal (Brikama, Pirang, Tendaba, Soma, Basse)	Q

Banjul City

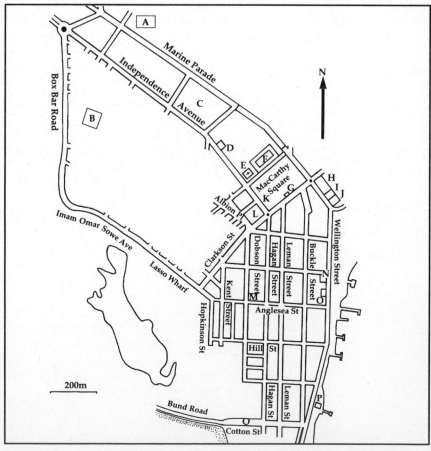

Food

There are several good restaurants in the tourist areas providing a variety of cuisines at a wide range of prices. Eating out in general is not expensive and a substantial meal at a beach bar will only cost D30-40. At the top end of the market, Yvonne Class Restaurant at Cape Point is the smartest in The Gambia and a meal for two with drinks may cost D800 or more, although this includes transport to and from the restaurant. Somewhat less extravagant is the 'Rive Gauche' á la carte restaurant at the Kombo Beach Novotel which offers French-style cuisine at affordable prices. Francisco's at Fajara also has a good reputation locally. The Sir William Restaurant, next to the Bakotu Hotel, is relatively inexpensive and has some Gambian-style dishes on the menu, including ladyfish and barracuda which are both worth trying. For a touch of the exotic visit the Bamboo Chinese Restaurant at Fajara or the nearby Siam Garden for Thai cuisine. The latter is especially good on Monday nights when they do an 'eat as much as you like' Thai buffet for about D120 (booking is essential).

Two Gambian specialities not to be missed are benachin (rice cooked in peanut or palm oil with vegetables and meat or fish) and domada (a fairly fiery dish of meat and vegetables cooked in a peanut sauce with a generous amount of chilli). Seafood in The Gambia is cheap and the large prawns are excellent, but avoid the oysters as they are a likely source of Hepatitis A.

In the Kotu/Fajara area, the Paradise Beach Bar, on the shore between the Bungalow Beach Hotel and the Fajara, is clean, friendly and reasonably priced, providing both snacks and more substantial meals (the prawns Gambian-style are excellent) together with what is probably the coldest beer in the area.

After a visit to Bijilo Forest Park, there are several small bar-restaurants on the beach between the Senegambia Beach Hotel and the Palma Rima. At lunchtime in Banjul, the Scandinavian owned and managed African Heritage Restaurant and Gallery, between the CFAO supermarket and the ferry terminal, is relatively inexpensive. It also serves morning coffee and is not far from the Banjul end of the Bund Road - unfortunately it is not open on Sundays.

There is a beach bar on the shore at the Fajara end of Atlantic Road; the track leading down to it is almost opposite the end of Pipeline Road. When birding on the Old Cape Road a number of establishments in the Cape Point area are within easy walking distance.

In any of the other birdwatching areas described, it will not be possible to find anywhere suitable to eat at lunchtime. Making a packed lunch is the only alternative to going hungry and when buying the ingredients try to avoid the 'mini-markets' near the tourist hotels as they are invariably expensive with a limited choice of goods. There are large CFAO (Compagnie Francoise de l'Afrique Occidentale) supermarkets on the corner of Wellington Street and Picton Street in Banjul, and almost opposite the African Village Hotel on Atlantic Avenue, Bakau. These stock Tesco's own brand produce as well as a good range of other well-known brands. There is also a small but good supermarket in Serekunda, next to the Standard Chartered Bank at the southern end of Kairaba Avenue. Tinned food is rather expensive, since it is all imported, but there is little practical alternative.

Fruit and salad vegetables are best purchased in local markets such as the ones at Serekunda and on Atlantic Avenue in Bakau, just before the turn off for Cape Point, where they are both cheap and of good quality. Fruit can also be bought from stalls on or near the tourist beaches, particularly next to the 'African Market' at Kotu, but the prices are much higher. All fruit and vegetables should be washed and peeled before consumption.

Gambian bread is excellent, cheap and delivered fresh to the shops several times a day. It contains no preservatives so is invariably dry and stale by the following day, but if a whole loaf is too large for you, just ask the shopkeeper to cut one in half - this is common practice and quite acceptable. Shops selling bread may be found in almost every street of every village and are usually open from early morning until late at night.

Some Sample Prices (1993):

Large beer	(beach bar)	D18
	(Kombo Beach Novotel)	D30
Soft drink - bottled	(shop)	D2.50
	(beach bar)	D6
	(hotel)	D10-15
Mineral water - large bottle	(supermarket)	D9
Gambian style prawns	(beach bar)	D30
Bacon, egg & salad roll	(beach bar)	D18
Bread - large loaf		D2
Eggs - each		D1.50
Corned beef - tinned		D30-40
Lemons - per kilo		D10
Oranges - each		D0.50
Grapefruit - each		D2
Water melon - per kilo		D5

Banks

The Standard Chartered Bank has its head office in Buckle Street, Banjul with branches in Kairaba Avenue, Serekunda, in Atlantic Road, Bakau (not far from the African Village Hotel), near the Senegambia Hotel in Kololi, and at Basse.

The Gambia Commercial & Development Bank has its main office in Leman Street, Banjul, with branches in Bakau (near the CFAO supermarket), Serekunda, Farafenni, Basse and at the airport.

The Banque Internationale du Commerce et l'Industrie is based in Senegal and has its Gambian branches in Wellington Street, Banjul, and in Bakau (behind the CFAO) and Serekunda.

The Central Bank of The Gambia issues the currency and has only one office, in Buckle Street, Banjul.

Bank opening times vary between individual branches and appear to change without warning, so it is best to check. Some banks will cash personal sterling cheques for a small fee, but credit cards cannot be used to obtain cash and are rarely of any use at all outside the tourist hotels.

Note that, despite The Gambia being a Moslem country, Sunday is the official day of rest and banks are open on Fridays. Public holidays are New Year's Day, Independence Day (February 18th), Good Friday, Labour Day (May 1st), Feast of the Assumption of St Mary (August 15th), Christmas Day, and a number of Moslem feast days which alter each year according to the Moslem Calendar.

Post and telecommunications

Stamps for postcards and letters can be bought from hotel receptions. They are also available from the beach vendors of postcards but only if cards are bought as well. Cards from these vendors are usually cheaper than those on sale in hotels and tourist markets. There are post offices in Russell Street, Banjul (the General Post Office) and on Kairaba Avenue, Serekunda. These are open from 08.30 to 16.00 on weekdays (closed for lunch 12.15 until 14.00) and from 08.30 until 12.00 on Saturdays. When posting cards or letters home it is best not to use the post offices. The post boxes in the hotel reception areas are emptied regularly and the mail is sent with the tour reps to the airport. Since different holiday companies have flights home on different days, it can be worthwhile checking with the various hotels in your area when their mail goes to the airport before posting anything.

The telecommunications system is operated by Gamtel, which has a number of offices where both local and international calls can be made from private booths. Gamtel offices are situated next door to the Kombo Beach Novotel at Kotu, just up the road from the Senegambia Beach Hotel in Kololi, near the CFAO supermarket in Bakau, in Telegraph Road (near the Atlantic Hotel) in Banjul, at the southern end of Kairaba Avenue in Serekunda, and at the airport. There are also call boxes near most of the hotels, some of which take coins (25 and 50 bututs only) and some phone cards, but at night these are sometimes in pitch darkness making a torch essential. A call to the UK costs D50 for two minutes, D45 if a phone card is used. Phone cards are on sale at Gamtel offices, but paradoxically cannot be used in them - only in those phone boxes equipped to accept them. Gamtel offices are open from 08.00 (or as soon after this as the operator turns up) until 22.00. Cheap calls operate after 18.00 on weekdays and all day on Sundays. Bear in mind that calls made from your hotel are likely to be considerably more expensive at all times.

To phone overseas from a Gamtel office, first dial 000 followed by the country code. For the UK, dial 000 then 44, then the STD code omitting the initial 0, and finally the number required. Thus 0891 700249 becomes 00044 891 700249. If dialling from a phone booth, omit the first zero (e.g. 0044 891 700249).

Short local calls are cheaper from a phone booth, where the minimum charge is 25b as opposed to the D1 at a Gamtel office, but always have a good supply of 25b and 50b coins. Unused coins are returned when the receiver is replaced. To make a local call from a Gamtel office, dial 0 before the number required.

To phone a Gambian number from the UK, dial 010 22 0 followed by the number required.

Electricity

The mains supply is 220-240 volts AC with some sockets being standard UK type 3-pin and others continental 2-pin, so a multi-adaptor is invaluable. Power can go off suddenly for no apparent reason, so most hotels have their own emergency generator and everyone else has a supply of candles.

'Give me pen'

Education, except nursery education, is free, but is only available to those children whose parents can provide them with the necessary equipment, which includes a desk, a chair, paper and a pen or pencil; hence the constant request, "Give me pen!" Unfortunately this has become almost a ritual demand and, certainly in the tourist areas, it is extremely doubtful whether the recipient of the gift will be scribbling away furiously with it the following day. More likely than not it will be sold to someone else.

It is far better, if you wish to take out some pens and notebooks to distribute, to give them to a teacher who will ensure that they reach the most deserving cases. Such gifts are especially welcome at the more out of the way village schools where tourists are a rarity.

Incidentally, the term 'toubab', which is used everywhere, and is especially called out by children, simply means 'white person' and is in no way intended to be derogatory or insulting.

'Bird guides'

Birdwatchers in the Kotu area will inevitably encounter the 'bird guides', especially at the Lower Bridge over Kotu Creek, between the Badala Park and Kotu Strand hotels. This is the haunt of a number of local 'bird guides' who are usually in attendance from about 07.00 until 09.30 and again in the late afternoon and early evening, looking for business. There are one or two 'licensed birdwatchers', identified by an official tag bearing their name and photograph, and they are knowledgeable and could be useful if you are a bit wary of rushing off on your own in a bush taxi. Lamin Sidibeh and Yusufa are two especially pleasant and friendly guides who can be trusted to look after you. The former can be contacted in advance c/o Sani Ceesay, Customs and Excise Department, Banjul, The Gambia. Typical charges are in the region of D140-D160 for a full day's guided birdwatching.

The unlicensed 'guides', without tags and identified solely by the pair of binoculars hanging round their necks, are a different matter. Their knowledge of birds is often superficial and it is far better, and cheaper, to birdwatch without them.

In December 1992, a number of enterprising young men began selling 'birdwatching tickets' for D5 each to birders on the Fajara Golf Course. The tickets were home-made but bore an official looking stamp. There is no charge for birdwatching on the Golf Course, or indeed anywhere else in The Gambia except in the reserves at Abuko and Bijilo, where properly printed tickets are sold from proper ticket offices, and at the Kotu Ponds where the 'ticket office' is rather makeshift but the tickets are genuine enough. Do not be tricked into paying for fake tickets.

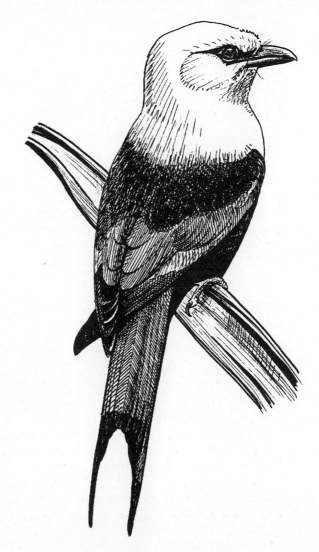

Blue-bellied Roller

CLIMATE AND CLOTHING

Lying, as it does, almost mid-way between the Tropic of Cancer and the Equator, and on the same latitude as Madras, The Gambia has a tropical climate of which the most significant feature is the occurrence of two markedly different seasons - the wet season which extends roughly from June to September, and the dry season from October to May. There is no sharp division between these seasons; rain still falls during October, though infrequently, and can sometimes start as early as May, especially up-river where the rains tend to start and finish rather earlier. Even during the wet season there are fine, sunny days, and most of the rain falls at night; humidity, however, is high and strenuous activity is definitely not recommended.

After the beginning of November the humidity falls rapidly, but even during the dry season there may be cloudy days and occasional light showers, especially during the latter part of December. There is a considerable difference between the climate inland and that on the coast, where there is usually a cooling breeze which keeps conditions bearable even during the hottest months.

Owing to the proximity of the country to the Equator, day length in The Gambia alters little throughout the year. Sunrise varies from about 06.45 in May to 07.30 in January and sunset from about 19.30 in July to 18.45 in November. Local time is GMT throughout the year.

Wire-tailed
Swallow

Cotton clothes are essential, including underwear and socks, and should be loose fitting. Avoid cotton-polyester mixtures as these do not 'breathe' as well as the pure material. Some people feel they need a thin jacket or pullover for the winter nights, which can be cool, and locals go around muffled in heavy anoraks and sweaters if the temperature drops much below 70°F, but anything more than a long-sleeved cotton shirt is rarely necessary. A long-sleeved shirt is also advisable if you intend to go looking for owls or nightjars at dusk, but this is for protection against biting insects, particularly mosquitoes, which can be a serious irritation (and potentially dangerous) once the sun goes down. Shorts are fine for the beach or the hotel gardens but elsewhere they are not a good idea. Apart from the offence which they may cause, particularly to older Moslems, shorts are associated, to Gambian eyes, with small boys and 'beach-bums'. Anyone wearing them, especially outside the tourist areas, will be the focus of attention and, if not indignation, at least considerable amusement. Apart from this, bare legs in the bush are an open invitation to biting insects and sharp thorns. Similarly, open-toed shoes or sandals, especially if worn without socks, offer little protection against the variety of spiny fruits and seeds which seem to take a vindictive delight in seeking out even the smallest area of exposed skin. A pair of tough suede boots which lace up to just above the ankles are recommended as they are comfortable to wear even in very hot conditions, offer good protection against spines and thorns, and some protection in the unlikely event of the wearer stepping on a snake. However, they are a bulky item to pack and a pair of comfortable trainers is an adequate alternative.

As the midday sun can be intensely hot, some form of hat, preferably with a wide brim to shade the back of the neck, is essential for anyone who is not accustomed to such conditions. In many areas this period is very quiet for birds, with early morning and late afternoon being the most productive times, but in the open bush it is the ideal time for raptor spotting as they begin to soar on the thermals.

HEALTH AND MEDICAL FACILITIES

Water   The standard advice from tour operators is to avoid tap water, and ice made from it, at all costs and stick to bottled mineral water. However, the water in the tourist areas comes from underground sources and is purified before distribution. Bottled water is quite expensive, especially in the hotels. I drank tap water with no ill-effects but it can be worth getting a couple of bottles of mineral water to start off with, if only to refill the empties from the tap. The water supply is unpredictable and it is as well to have a reserve on hand.

Up country it is best to buy mineral water, or take a supply. Several litres of water need to be drunk per day, especially if out birdwatching in the often intense heat. Mineral water bottles are awkward to carry and easily punctured - a good sturdy water-bottle on a strap, with at least one litre capacity, is much more useful and can be filled with tap water, or mineral water if preferred. Warm water can taste horrible and a good tip is to squeeze some lemon juice into the bottle which helps to keep it palatable.

Insects

Biting insects, including mosquitoes, start to make their presence felt at dusk, so for any evening excursion you should go amply supplied with insect repellent. The various brands differ in their effectiveness, but those containing DEET (diethyl toluamide) seem to be the best.

Sunburn

This is painful and potentially dangerous. Beware of hazy days when the sun does not feel especially hot, and use plenty of high factor sun cream, at least for the first few days.

'Banjul Belly'

Most visitors are struck down by this eventually, some briefly and annoyingly, others more seriously. The cause is a matter for speculation and there are a number of theories. The best treatment is to starve for twenty four hours and take nothing but lemon tea (lemons are readily available from local fruit sellers). Kaolin and morphine helps ease the often painful stomach cramps but otherwise it is probably best to let the bug run its course, bearing in mind that there is no shortage of bushes in the bush! The main danger is dehydration, and this can be avoided by drinking plenty of water while making sure that salt intake is adjusted to compensate for loss. If dehydration symptoms do occur (thirst, dizziness, lethargy and sometimes nausea and muscle cramps) a simple remedy is to dissolve a teaspoonful of salt and four teaspoons of sugar in a litre of water and sip the liquid slowly. Those who are suceptible to stomach upsets should consider bringing Imodium tablets and Dioralyte rehydration powders which can both be bought from a high street chemist without a prescription.

Anyone who is thinking of spending any length of time in The Gambia, or contemplating 'roughing it' in non-tourist hotels and rest houses up-river, can obtain more detailed medical advice from MASTA (Medical Advisory Services for Travellers Abroad Ltd), Keppel Street, London WC1E 7HT. A small fee is charged for the service.

The larger hotels have a nurse or doctor on call for minor problems and will contact the nearest hospital in an emergency. Anyone on a package tour will have a tour company rep to sort things out. The main hospital is the Royal Victoria Hospital on Independence Avenue, Banjul, which has well-qualified staff and good facilities. There are also well-equipped clinics at Kanifing, and Kololi. Treatment is not free, however, and adequate medical insurance is essential.

MAPS

Good maps of the country are available from Edward Stanford Ltd in London, the most useful being the large scale (1:50,000) sectional maps produced from aerial photographs by the Directorate of Overseas Surveys. These are somewhat out of date but the main roads and most of the tracks remain unchanged, although some of the minor tracks and paths shown may bear little resemblance to those actually occurring on the ground, since it does not take long for a new bush track to become established or for an unused one to disappear. The most useful map for the coastal area is Sheet 10 (Banjul). References given in the text are based on these maps.

Also available from Stanford's is a 1:300,000 tourist map which includes an information booklet, and a 1:250,000 DOS map which shows administrative divisions, communication networks and national parks. The Gambia National Tourist Office provides, free of charge, a useful small-scale tourist map of the country, together with other helpful information.

Useful addresses are: Edward Stanford Ltd., 12 - 14 Long Acre, London WC2E 9LP (Tel. 071 836 1321); National Tourist Office, The Gambian High Commission, 57 Kensington Court, London W8 5DG (Tel. 071 937 6316).

Rufous-crowned Roller

WHEN TO GO

The rainy season lasts from June until the end of September, although the rains may occasionally begin as early as May and sometimes extend into October. Birdwatching is rewarding at any time but many changes occur during the dry season and these are summarized below.

October: Although the rains have finished there is still the chance of a shower, rarely prolonged or heavy. Cheap holiday package deals are available this month and would suit people on a budget. The weather can be hot and humid, but a breeze off the sea usually keeps conditions bearable, at least in the coastal areas. The countryside is green and butterflies abound. Small pools of water by the roadsides and elsewhere attract large numbers of smaller birds and provide great photographic opportunities. Resident species are in full breeding plumage, but the tall grasses which occur in many areas impede visibility. This is a good month for passage migrants, which are at the height of their southern migration; most continue their journey but some, including Nightingales and a number of warblers, are commonly seen throughout the winter. Sea-watching can be rewarding, with Black, Common, Royal, Caspian and Sandwich Terns regularly cruising past.

November: Temperature and humidity fall steadily during the month, and by the final week the climate is very pleasant, with hot, dry days and cool mornings and evenings. The countryside is still green for the first part of the month, but by the end the more open areas are starting to look parched, the smaller pools are dry and even the large waterholes are shrinking noticeably. Bishops and whydahs are still in breeding plumage but will soon start to look rather tatty. Long grass is still much in evidence, giving the same problems as October. Mid-month brings dramatic changes, with tern numbers falling markedly and other species becoming common as the winter and dry-season visitors begin to arrive in force. Larger birds are obviously more noticeable and suddenly Black Kites, Grasshopper Buzzards and Abyssinian Rollers seem to be everywhere, with Black-shouldered Kites also appearing in large numbers in some years. This also seems a particularly good month for Western Marsh Harriers. For anyone not on a tight budget and preferring more comfortable climatic conditions, the middle two weeks of November are perhaps the ideal time for a birdwatching trip to The Gambia.

December: The countryside continues to dry out rapidly, with only the largest waterholes still holding an appreciable amount of water, so that wildlife tends to become concentrated to some degree. Cereal crops are being harvested and so more open areas are appearing, although the tall grasses remain and continue to restrict visibility. Some 'burning-off' occurs, around Kotu Ponds for example, and there is a good chance of spotting Temminck's Coursers in the water melon fields around Lamin and Yundum. Bishops and whydahs have now become difficult to identify with males looking as drab and inconspicuous as

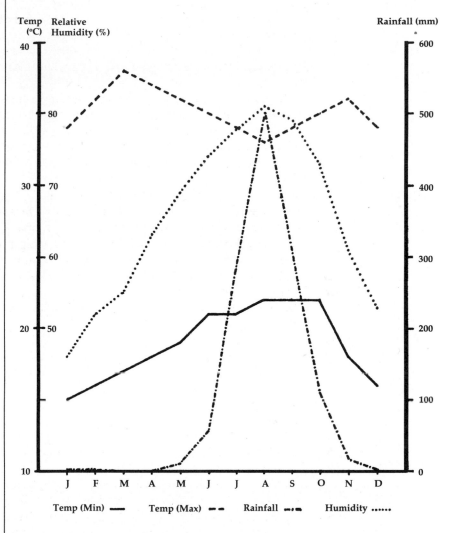

*Temperature,
rainfall and
humidity*

the females. By mid-month, the cereal harvest is complete inland, although lagging behind nearer the coast, and the fields are rapidly being cleared, which means that ground-feeding birds, including Abyssinian Ground Hornbills, are suddenly much more visible.

Acacias are in flower, and attract large numbers of sunbirds, warblers and estrildine weavers, although these may be difficult to see amongst the dense foliage. Expect some cool, overcast days, with even the chance of a shower later in the month. Although this makes the mid-day period more bearable, it leads to a dearth of soaring raptors because of the absence of thermals, and even small passerines seem reluctant to come out of hiding when the sun is not shining.

January: The New Year may bring some cloudy days and the chance of a light shower. Evenings are cool and early mornings may be chilly, but when the clouds disperse, and the sun breaks through, the days can still be very hot indeed. January is the month when the 'Harmattan' may blow down from the Sahara. This is a hot dry wind which may reach gale force at times, filling the air with red dust and making birdwatching, and any other outdoor activity, virtually impossible. The Harmattan used only to occur once every few years; recently the frequency has increased but, even so, you would be unlucky to encounter the full-blown variety.

There is little change on the bird front near the coast, but the Egyptian Plovers will have gone from Basse by the end of the month. The countryside continues to dry out rapidly, but has not yet taken on the really parched look which it will have acquired by late March and April. Acacias finish flowering, but in the latter part of the month the Red Silk-cottons begin to come spectacularly into bloom, attracting many birds which, since the trees lose their leaves as they come into flower, are quite easily observed.

February: Temperatures continue to rise and humidity remains low, although there is still an outside chance of a brief shower. The last of the grain crops have been harvested by now, the fields have been cleared and more burning-off occurs, so that the countryside takes on an even more open and sometimes barren look. Burning areas can be rewarding for birds, as the smoke often attracts numbers of rollers, hirundines and raptors, which feed on the insects, lizards and small mammals as they flee from the flames. February sees the beginning of the spring passage northwards, although this is less impressive than the southern movement and will not be in full swing for another month.

March: This is often the hottest and is certainly, with April, the driest time of the year. Grass fires have left much of the savannah looking black and unattractive but, once the tall elephant grass has gone, movement is easier and the birds are much more visible. March sees the peak of the northern passage with terns appearing in large numbers again, including White-winged Black Terns, which are not seen in The Gambia on their southerly migration. European Turtle Doves may also appear in large numbers in March.

April: Hot and dry, although still with cooling breezes on the coast. April sees the main tourist season drawing to an end, with many hotels closing down at the end of the month. Northward-bound passage migrants remain much in evidence, although the peak period has now passed. White-winged Black Terns are still abundantly present and by the end of the month are coming strikingly into breeding plumage. A trip across to Barra on the ferry during the last week of April may be result in the sight of several hundreds on the water off Banjul. Weavers too are starting to come back into breeding plumage and, since many of the dry season visitors are still in residence, the latter part of April can be a rewarding time in some areas. Visitors to the up-river sites may be disappointed since the dried-out state of the country means far fewer species are likely to be seen than at the beginning of the season.

INTRODUCTION TO THE SITE INFORMATION

Almost anywhere in The Gambia will provide interesting birdwatching and it is impractical to attempt to cover every possible location within the confines of a guide such as this. The sites described have been chosen because they provide examples of a wide range of habitats and thus the opportunity of seeing as wide a variety of species as possible. They will, as an added bonus, provide an experience of the country which is denied the casual visitor. Gallery forest, savannah woodlands, mangroves and rice fields, estuarine creeks, coastal lagoons and village cultivations can all be visited, as well as the remarkable Golf Course and the gardens at Fajara.

Because of the problems of long distance travel and up-river accommodation, the majority of sites described have been chosen so that they are fairly easily visited in a day trip from the tourist areas, and some of them are within a short walk of many of the hotels. Only Tendaba and Basse will require an overnight stay and accommodation for these is dealt with in the appropriate site description.

The birds described for the various sites are not an exhaustive list, but are those which are particularly characteristic of the site or which are likely to be seen at few, if any, other localities. However do not be surprised by anything, avian or otherwise, as rarities can turn up in the most unlikely places, while the commonest species may suddenly and unaccountably become conspicuous by their absence. A full list of all the species reliably recorded in the country is provided at the end of the guide.

The sketch maps are meant to provide an easy-to-follow guide around the various sites and are not necessarily intended to be cartographically precise. The direction of north is given as a guide only and should not be used to set a compass bearing.

Map references relate to the 1:50,000 sectional maps available from Edward Stanford. The sheet number is given first, followed by a six figure reference.

Reference is sometimes made to Lower, Middle and Upper River regions of the country. As a guide, the Lower River region may be taken to extend from Banjul eastwards to Tendaba, the Middle River from here to Georgetown, and the Upper River from Georgetown to the eastern boundary of the country.

Abuko Nature Reserve

The Abuko Nature Reserve was established in 1967. The area enclosed, although relatively small, includes a fine example of mature gallery or riverine forest through which runs the Lamin Stream. Such primary forest is now very rare in The Gambia due to clearance for agriculture, and Abuko provides the most easily accessible example in the country. It has remained undisturbed as a water catchment area for many years, and ground water is still extracted by the pumping station near the main road. Surrounding the central forest is more open savannah woodland, and the trail through the reserve passes through both types of habitat.

The official bird checklist, as revised in 1984, comprises 201 species and has doubtless increased in the meantime. Among Abuko's scarcer specialities are Western Little Sparrowhawk, African Goshawk, White-spotted Flufftail, Guinea Turaco, Verreaux's Eagle Owl, Lemon-rumped Tinker-bird, Spotted Honeyguide, Buff-spotted Woodpecker, Grey-headed Bristlebill, Yellow-breasted Apalis, Green Hylia, Collared Sunbird and Western Bluebill, although the chances of seeing all, or even most, of these in a single visit are extremely remote.

Location

Map Reference (Entrance) - Sh10:218810. The reserve lies on the main Banjul-Basse highway, just before the village of Lamin and about 7km from Serekunda. The entrance to the reserve, which is clearly signposted, is just after the Abuko weighbridge.

Tourist taxis operate fixed-price excursions which give two or three hours visiting time, but this is inadequate for anyone who wishes to see the reserve properly. It is often possible, however, to come to a private arrangement with a taxi driver by which he will drop you at the reserve and collect you again at a pre-arranged time, allowing a longer and more profitable visit. This can be expensive; a price should be agreed before embarking on the journey, ensuring that no money changes hands until the return trip.

Buses from the GPTC terminus in Half Die (Banjul) travel via Kanifing and Serekunda before passing the reserve on their way up river to Soma, but they are often full and rather unpredictable.

Minibus bush taxis offer the cheapest and most convenient alternative, as they leave Serekunda for Brikama at frequent intervals and there should be no problem stopping one for the return trip - a shady tree on the opposite side of the road and just down from the reserve entrance is a convenient place to wait.

Strategy

It is easy to spend a full day in the reserve, permitting both early morning and late afternoon visits to the Crocodile Pool where different species appear from time to time. Even during the mid-day period there is usually something to be seen. Refreshments are available from a kiosk in the 'Animal Orphanage' but these are limited to soft drinks and rather dry cake. There are, however, several benches in the reserve where a packed lunch can be eaten in relative comfort.

The admission charge was D15 in 1993 and the opening hours 08.00 until 18.00 (closed public holidays). Guide books are available, both to the trees and shrubs (a selection of which have numbers painted on them to aid identification) and to the reserve in general. The latter is really only a glossy souvenir booklet but contains some nice bird photographs as well as a map.

Anyone hoping to use the photo hide at the 'Animal Orphanage' must book well in advance, since occupancy is restricted to a maximum of three and it is sometimes block-booked by companies offering birdwatching trips.

Birds

The route through the reserve is clearly indicated and there are numbered, green, metal marker posts at twenty metre intervals (although one or two of these are missing). The first stretch of the path passes through savannah woodland with a variety of trees and shrubs before arriving at the north-east end of the Crocodile Pool, crossing a small inlet (dry during much of the season) by means of a metal bridge. Here the path enters the gallery forest proper and a little further on is the Crocodile Pool Education Centre which has a convenient viewing verandah overlooking the pool. It is worth spending some time here at both the beginning and the end of a visit since species tend to come and go, although Black-crowned Night-Heron, Black-headed Heron, and Hamerkop are usually in evidence, while Fanti Saw-wings often circle over the water and Palm-nut Vultures frequent the Oil Palms beyond the far end. The Black-crowned Night-Herons roost to the right of the verandah in the Gumbar trees which were planted on the opposite bank to hide the pumping station, and also in the Raffia Palms on the far side. There is also a chance of White-backed Night-Heron here, though this is more likely early in the season. Other species which regularly visit the pool-side include Black Heron, Black Crake and Giant Kingfisher, while various raptors appear from time to time and the elusive African Crake may put in an appearance. The trees and bushes to the left of the verandah should not be ignored, since Collared Sunbirds are sometimes seen there.

Just beyond the Education Centre, a path leads down to the right to a public photo-hide which overlooks the south-west end of the pool. African Darters are common at this end while African Jacanas also occur regularly and there is the chance of a Western Little Sparrowhawk. White-spotted Flufftail is a rare resident in the reserve and never ventures far from water, but is one of the shyest of birds and rarely observed.

Beyond the path to the hide, the main route passes through dense gallery forest where several species may be seen among the branches, including African Paradise Flycatcher, Brown-throated Wattle-eye and Buff- spotted Woodpecker. The path continues past some impressively large Grey Plums after which, at marker post number 24, another side path leads to a second public photo-hide, not as well sited as the first but worth a quick visit. The hide is on two levels but the upper story is difficult to enter with ease and leaving it is even more tricky.

At marker post number 29 a cross path leads through a depression, wet during the period immediately following the rainy season, and

Key

Abuko Weighbridge	A
Entrance and Ticket Office	B
Education Centre and Observation Verandah	C
Toilets	D
Crocodile Pool	E
Public Hides	F
'Animal Orphanage' and refreshments	G
Private Photo-hide (pre-bookable)	H
Antelope and Crowned-Crane Enclosure	I
Hyenas	J
Tree-top Hide (closed)	K
Exit and Mini Craft-market	M
Pumping Station	N
Wildlife Conservation Department (private)	P

The numbers are those of the 20m marker posts

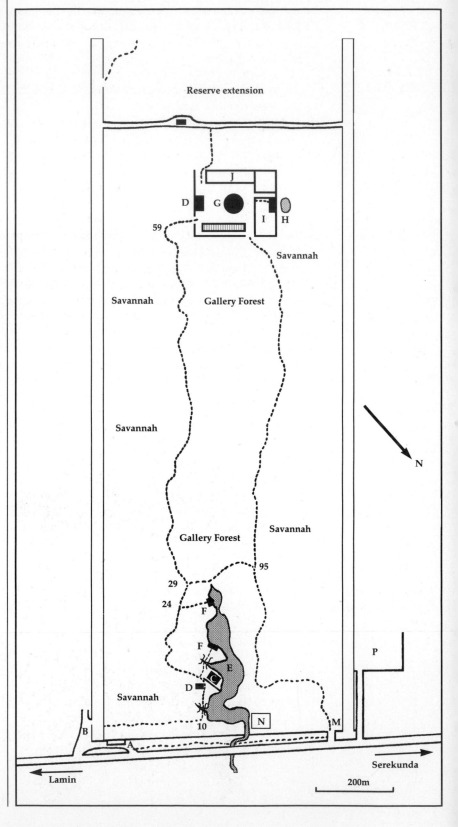

connects with the path returning on the other side of the reserve. It is best to ignore this for now and continue through the forest until, near post 34, it begins to open out with good views over clearings to the right where a variety of sunbirds and warblers are usually present. It is also worth keeping an eye on the tree-tops from time to time since Violet Turacos, Guinea Turacos and African Pied Hornbills seem particularly to favour this area.

At the south-east end of the path the 'Animal Orphanage' is entered, with a few sad-looking parrots in cages as well as hyaenas, a pair of lions, Black Crowned-Cranes and a selection of antelopes. Although it is stressed that these are all creatures which, for one reason or another, cannot be returned to the wild, and the area is not a zoo, it cannot help but give the impression of being one.

A path leads down by the side of the hyaena enclosure (which is usually teeming with opportunist Hooded Vultures and extremely foul after feeding time) and shortly reaches a track which runs across the end of the original reserve. Beyond this lies the Abuko extension, an area of savannah through which runs an 'ornithological walk'. The extension is best visited in late afternoon and although most of the species there may have already been encountered elsewhere, a visit should be included if time permits as there is a good chance of seeing some of the less common raptors for which Abuko is noted. Long-tailed and Standard-winged Nightjars also occur here at dusk, but special

African Paradise-
Flycatcher

permission will have to be sought at the ticket office to stay on after the usual closing time.

If you have booked the private photo-hide, then the attendant at the Animal Orphanage will unlock the gate to the antelope enclosure and escort you through it to the hide. This overlooks a small artificial pond, constructed from concrete, which is hardly aesthetically pleasing but does attract a number of species within telephoto range.

The return path from the Animal Orphanage leads down the north-west side of the reserve, with the forest on the right and rather more open savannah woodland on the left. At marker post 95 the cross path from number 29 comes in from the right, and this provides a convenient way to return to the entrance. The main path continues through mainly savannah to the exit but is not especially rewarding and there is a small craft market just before the exit gate with stall-holders who can be irritatingly persistent. The exit is about 400m from the entrance, along either the main road or a not-very-good path which entails jumping across a conduit at one point. If you have arranged with a taxi to pick you up, it is important also to arrange in advance whether it will wait at the exit or the entrance.

The cross-path, though short, has some good birdwatching points and should not be hurried. It leads back into the dense gallery forest just before rejoining the main path, where a left turn will bring the visitor back to the Crocodile Pool, providing the opportunity for a final look before leaving.

Other wildlife   Several species of mammals and reptiles inhabit the reserve, but many are nocturnal or crepuscular and are unlikely to be seen by the casual visitor. Those most likely to be encountered include the Green Vervet and Western Red Colobus Monkeys, with a chance of Patas in the more open areas of savannah (one which habitually lurks around the 'Animal Orphanage' is semi-tame). Two other common mammals are the Gambian Sun Squirrel, often seen climbing around in the trees, and the Striped Ground Squirrel. The only reptiles likely to be seen are the Common Agama lizard, the Nile Monitor, Bosc's Monitor and the Nile Crocodile. The first three are completely harmless and extremely shy while the fourth is confined safely to the immediate surroundings of the Crocodile Pool, which is also sometimes called the Bamboo Pool, 'bamboo' being, somewhat confusingly, the Mandinka word for crocodile!

Several species of snake are present, including a number of venomous varieties such as the Green Mamba and the Black Cobra, but very few visitors ever see one and no-one has been bitten by a snake in the entire history of the reserve. To avoid being the first do not stray from the clearly defined paths, which in any case may disturb animals and plants alike.

One impressive creature which may be encountered early in the season is the female Nephila spider - up to three inches across and with a handsome yellow-banded body and black legs which shine iridescent purple in a good light. These are completely harmless, unless you happen to be a male Nephila!

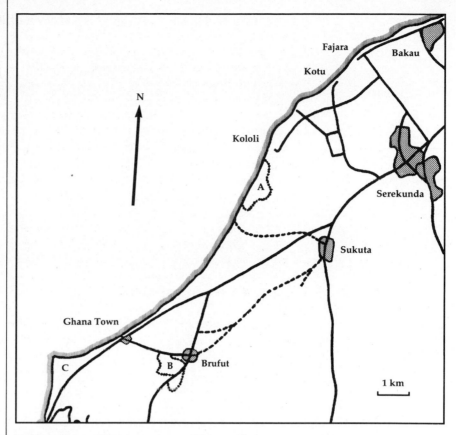

Bijilo Forest Park

Bijilo is a relatively new forest park nature trail, only opened to the public in 1991, which comprises an area of Rhun Palm forest and coastal scrub on a sloping site at Kololi. It is both picturesque and interesting and is well worth a visit, not only for the birdlife but also because it is one of the last natural Rhun Palm stands of any size which remain in The Gambia and is in complete contrast to the gallery forest of Abuko. Among the species which regularly occur are: Osprey, Stone Partridge, Violet Turaco, Grey-headed Kingfisher, Swallow-tailed Bee-eater, Rufous-crowned Roller, Black Scimitarbill, Vieillot's Barbet, Cardinal Woodpecker, Oriole Warbler, Northern Crombec, Senegal Batis and Green-headed Sunbird.

Location   Map Reference (Entrance): Sh10:132861. The reserve lies close to the Kairaba and Senegambia Beach Hotels and the Kololi Beach Club, and is reached along a wide laterite track which branches off the road leading down to the hotels immediately opposite the Standard Chartered Bank and adjacent to the Kololi bush taxi depot. After a few yards the track passes the entrance to the Kololi Beach Club and then shortly bears sharp right at the entrance to the trypanosomiasis research centre. The entrance to the forest park is just down the hill on the left hand side and is impossible to miss.

From the Kairaba and Senegambia Beach Hotels the reserve is only about ten minutes away on foot. From the Palma Rima it can be reached along the beach in about twenty minutes (a more pleasant walk than along the road), and from the hotels in the main Kotu area in about three quarters of an hour.

Bush taxis run from Serekunda to the depot at Kololi fairly frequently and pass the Palma Rima. From the hotels along Atlantic Avenue and at Cape Point it will be necessary to pick up a taxi in Bakau for Serekunda and then walk down the road to board a second one for Kololi.

Strategy   It is possible to spend a full morning or afternoon in the reserve or walk round it in just a couple of hours but, as at Abuko, the more time invested the more will be seen.

Early morning and late afternoon are the best times to visit, but avoid overcast days, when little seems to move and the reserve may appear totally lifeless. The 'ornithological path' is very exposed and the sun beats down on it fiercely during the early afternoon so wear a hat and take a bottle of water along since refreshments are only occasionally available.

The entrance fee is the same as Abuko (D15 in 1993) and the opening times are also the same: 08.00 until 18.00 and closed on public holidays. There is a guide book available, price D15, describing the vegetation of the reserve, but this contains no map or illustrations and is really only of interest to dedicated botanists. The nature trail comprises four, interconnecting, circular routes of varying lengths labelled green, red, blue and yellow. There is also an 'ornithological path' linking the green and yellow routes. All footpaths are marked with appropriately coloured posts along the edges and there are clear maps at every intersection, so it is impossible to get lost.

The arrangement of the paths allows considerable flexibility in the choice of a birdwatching route, which can be tailored to suit the time available. It is worth including the 'ornithological path' in any itinerary, since it provides good views across the more open areas of the reserve and passes through a variety of habitats. Combining this with complete coverage of the other routes means an early breakfast and a late lunch. It is possible to reduce the time required, and still see a good cross-section of the reserve, by following the green route as far as the ornithological path, birdwatching along this to its junction with the yellow route, and then returning by way of those sections of the other routes which run along the upper (eastern) side of the reserve.

Birds   From the entrance gate a single track leads into the reserve, passing, after a few yards, a short, narrow path which branches off to the right and leads only to a small clearing. Not far beyond this is the beginning of the green route, which branches left and right and has a total length of 1400m. A right turn here leads through tangled scrub covered, as in many areas of the reserve, with a dense mat of the hairy-stemmed climber Merremia aegyptiaca. There are several fairly open areas along this part of the path, with scattered Rhun Palms and broadleaves which include Gingerbread Plum, a very common shrub or small tree in coastal

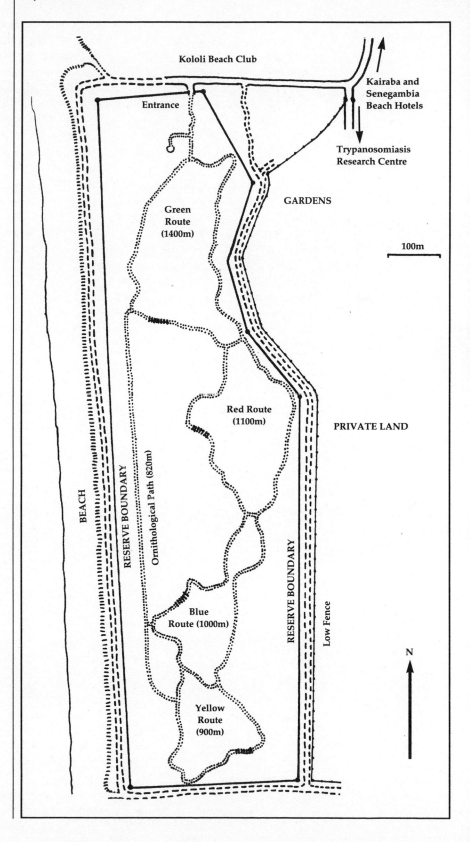

Kololi Beach Club

Entrance

Kairaba and
Senegambia
Beach Hotels

Trypanosomiasis
Research Centre

Green
Route
(1400m)

GARDENS

100m

Red Route
(1100m)

PRIVATE LAND

BEACH

RESERVE BOUNDARY

Ornithological Path (820m)

Blue
Route (1000m)

RESERVE BOUNDARY

Low Fence

Yellow
Route
(900m)

N

regions. It is worthwhile spending some time here, as it can be one of the most rewarding birding areas in the reserve, with the possibility of species such as Violet Turaco, Grey-headed Kingfisher, Swallow-tailed Bee-eater and Northern Crombec. The path opens up on the right hand side as it nears the sea, with low scrub and areas of grass and herbaceous plants dotted with a few Rhun Palms and stunted Baobabs. Soon the ornithological path branches off to the right in the midst of a small grove of Oil Palms, where a wooden bench is thoughtfully shaded by a palm-thatched shelter. This path, coloured black on the maps and the tops of the marker posts, is a more-or-less straight, sandy track running along the seaward side of the reserve. There are good views to the left over mixed scrub and low trees towards the Rhun Palm forest on the rising ground inland. Here Grey and Cardinal Woodpeckers nest in the older palms, Osprey and Palm-nut Vulture are regularly seen in the tree tops, and the lower scrub may hold Northern Black-Flycatcher, Black-necked Weaver and the occasional Black Scimitarbill. The wire fence which forms the seaward boundary of the reserve is a favourite perch of a number of birds, including Little Bee-eaters and Rufous-crowned Rollers, while the scrub between this and the path commonly holds Yellow-crowned Gonolek, Black-crowned Tchagra and the occasional Striped Kingfisher.

As well as birds on the reserve itself, a variety of gulls and terns may be seen flying over the sea at any time of year, but most especially during October and mid-November and again during March and April.

The ornithological path eventually joins up with the yellow route where there is a choice of either turning right and walking round the upper part of the circuit, or turning left and following the lower part. The second option is marginally shorter in distance, but the upper track provides the better birdwatching. It runs at first through a mixture of

Stone Partridge

open areas and dense, high scrub which often has plenty of audible bird life but generally poor visibility. Eventually the path turns inland, close to the boundary fence of the reserve and runs between dense stands of young Rhun Palms before a steep upward gradient leads to a flight of steps, at the top of which a conveniently sited and shaded seat offers the chance of a breather with good views over the tree tops and the possibility of Vieillot's Barbet or Green-headed Sunbird.

The path turns to the left at the top of the hill and runs along the upper, inland edge of the reserve passing through a variety of vegetation. Where young palms enclose the path in a green tunnel there is usually little to be seen, although it is worth walking cautiously since there may be Stone Partridges on the path, and these can vanish incredibly rapidly if disturbed. More open areas of scrub and mixed trees provide more varied birdlife, with the possibility of Blackcap Babbler, Oriole Warbler, Batis Flycatcher and Black-necked Weaver.

Following the eastern sections of the yellow, blue and red routes, the path finally joins the green route again. Here a right turn onto the upper section eventually completes the circuit at a 'T' junction, where the right hand path leads back to the entrance.

As well as the various paths through the reserve, there is a footpath which makes a complete circuit around the outside of the boundary fence and can provide some good birdwatching. The western section runs along the top of low, laterite cliffs and offers good views over the reserve as well as the shore and coastline. The south section passes through an area of Rhun Palms of various ages, while the eastern section is separated from the private grounds of the trypanosomiasis research centre by a low fence. The grounds are rather open and park-like to begin with, with just a few scattered trees, but towards the northern end the path passes alongside the flower-filled gardens of what appears to be staff accommodation, and these attract a variety of birds including warblers, sunbirds and estrildine weavers.

Other wildlife   Apart from the birds there are Green Vervet and Western Red Colobus Monkeys, Gambian Sun Squirrels and Common Agama lizards. There are also snakes including the handsome, black and yellow striped African Beauty Snake and the highly dangerous Puff Adder - always keep to the path where you can see clearly what you are about to step on.

Tanji Reserve

The Tanji bird reserve, established by professional ornithologist Clive Barlow, consists largely of fairly open areas with stands of Gingerbread Plum and other small trees, and a wide variety of other habitats including Acacia scrub, an area of young Rhun Palms, a shallow lake (which dries up rapidly after the end of the rainy season) and a sizeable wooded area containing mature trees which harbour Violet Turacos and other interesting species. The Green Crombec, a species only recently added to the Gambia list, has been seen here. In addition, the low, sandy cliffs which constitute the seaward boundary of the reserve offer good vantage points for viewing the Tanji lagoon, which attracts large numbers of terns and waders in spring and autumn, although these tend to be less numerous during the remainder of the dry season.

Location

Map Reference - Sh10:218810. The reserve lies between the Tanji - Tujering road and the Tanji lagoon, about 10km south-west of Sukuta and just to the north of the Tanji fish-curing site. There are no boundary fences and no restrictions on entry. The main access point lies about a mile and a half from Ghana Town, at the southern end of the mature, wooded area and just before the more open region of Gingerbread Plum which stretches from here to the mouth of the River Tanji. It is clearly marked by signs requesting visitors to refrain from any activity which might cause damage or disturbance to the plants and wildlife.

Bush taxis pass the reserve but are infrequent and usually full; the same is true of the GPTC buses. Although journeying to the reserve by one of these methods might not be too difficult, the return trip could prove a real problem.

Organised tours, some led by Clive Barlow, visit the reserve regularly and are convenient for anyone happy to watch birds with a dozen or so fellow enthusiasts. For greater independence, a tourist taxi offers the

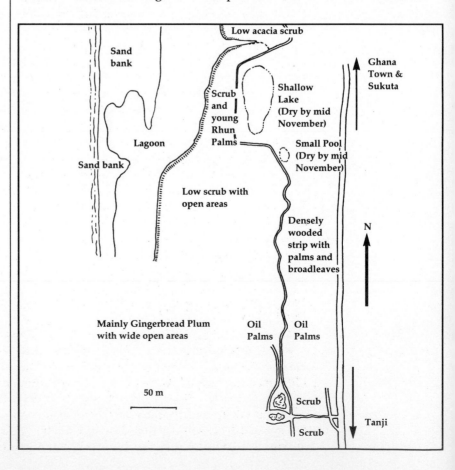

best solution. Prices vary slightly, depending on the distance from your hotel, but in 1993 were in the region of D240 for a return trip, including three hours waiting time. This should be quite sufficient to allow coverage of the most popular and rewarding areas of the reserve, and since the fare is per car, and not per person, it works out much cheaper than the organised excursions if two or more people go together.

With independent transport, if time permits, it can be worthwhile stopping on the way out of Ghana Town and birdwatching along the sides of the road near the wind-pump, where Brown-throated Wattle-eye, Greater Blue-eared Glossy Starling and a variety of other species may be present.

Strategy

The reserve is best visited in the early morning, preferably before 10.00, since after this time it can become very hot and many of the birds start to take cover. Late afternoon is also a possibility, but then views across the lagoon are into the sun. Spring and autumn are best for waders and especially for terns, when large mixed flocks appear, but the reserve provides excellent birdwatching at any time of year.

Birds

At the main access point, ignore a wide track off to the right and follow a narrower path which leads down a steep slope between dense bushes and trees where African Pygmy Kingfisher, Grey-headed and Sulphur-breasted Bushshrikes and African Golden Oriole may sometimes be seen.

The path leads down to an area dotted with widely separated clumps of Gingerbread Plums and other small trees, where Swallow-tailed Bee-eaters, White-crested Helmet-Shrikes, Northern Black-Flycatchers and several species of sunbird occur, with the chance of a Striped Kingfisher or a Black Scimitarbill. A well-defined track runs to the right along the edge of the mature wooded area, which occupies a low ridge and is a haunt of Osprey and Palm-nut Vulture as well as the magnificent Violet Turaco, while the bushes below may provide a Northern Puffback or, in the early part of the season, a Levaillant's Cuckoo.

The track continues through a small group of Oil Palms and winds its way towards a large and picturesque baobab tree, from which point it runs downhill towards a sizeable open area. Until about mid-November this is partly occupied by a shallow lake and attracts a variety of waders, while the surrounding scrub may hold Black-necked Weaver. A narrow path leads to the left and follows the western side of the depression. To the left is a low bank, covered with young Rhun Palms and various small shrubs, beyond which, out of sight, lies the lagoon.

There are several places where narrow paths lead up through the scrub to vantage points overlooking the shore, but beware of the so-called 'Welcome Grass' (Cenchrus biflorus), whose spiny fruits can cause very irritating and unpleasant sores if they manage to get inside shoes.

Beyond the lake area, the main track widens once again and turns right towards a region of dense Acacia scrub where Black-crowned Tchagras often sing. A small, dry creek, coming in from the left, allows

Violet Turaco

access to the shore which, when the tide is out, can be followed south along the edge of the lagoon to several points where the low cliffs can easily be climbed to give good views over the water and the sand-bank beyond. There is usually a variety of waders here, especially in spring and autumn, together with impressive numbers of Common, Royal, Sandwich and Caspian Terns, and the occasional Little, Lesser Crested or Gull-billed Terns. It is also a good place to see a Giant Kingfisher perched on a dead tree which protrudes picturesquely from the waters of the lagoon.

The thick bush bordering the lagoon is worth exploring for African Pygmy Kingfisher, Grey-headed Bushshrike and White-rumped Seedeater while any trees in fruit should be checked for African Green Pigeon, especially during January when the figs are fruiting.

It is preferable to return by the same route as far as the main area of Gingerbread Plums, where a roughly circular detour will conveniently bring you back to the path by which you entered.

Other Wildlife

Green Vervet and Western Red Colobus Monkeys occur in the more wooded areas. Around the edges of the lagoon there are often large numbers of fiddler crabs.

Brufut Woods

This relatively small area of undisturbed savannah woodland is one of the few still remaining in the coastal area. The woods harbour a variety of interesting species including Violet Turaco, Didric and Klaas's Cuckoos, White-faced and African Scops Owls, Long-tailed Nightjar, Mottled Spinetail, Vieillot's Barbet, Lesser Honeyguide, and Snowy-crowned Robin-Chat. Green Crombec has also been reported.

Location

Map Reference - Sh10:095795. The woods lie to the south-west of the village of Brufut, mainly in the angle formed by the junction of the Brufut-Ghana Town and Brufut-Tujering roads.

Bush taxis run from Serekunda to Sukuta where a change for Brufut will almost certainly be needed. The taxi depot in Sukuta is in the main square, by the market. The best area for birds is quite a walk from the main road, and anyone intending to stay until dusk for the owls may not enjoy the return trek through the woods in pitch darkness. The best solution is to hire a four-wheel drive vehicle and driver, which should not be very expensive for a trip lasting two or three hours, particularly if three or four people share the cost. It also avoids being followed by half the village children.

Most taxis reach Brufut along the laterite Ghana Town road out of Sukuta, branching off left for Brufut after about 5km, but if you have your own four-wheel drive vehicle and driver, it might be worthwhile taking the sandy bush track which runs direct from Sukuta to Brufut, as this can be good for small raptors; Grey Kestrels are especially common here and there is a good chance of a Dark Chanting Goshawk. A couple of miles out of Sukuta there is a distinct fork in the track; the right fork goes off to join the main laterite road, and it is best to take the left fork, which runs directly to the central crossroads in Brufut.

The woods are reached down a narrow track which leads off the Brufut-Ghana Town road. On reaching the central crossroads in the

Key

Village well	A
Village rubbish dump	B
Narrow, overgrown track – passable with 4WD	C
Parking space	D

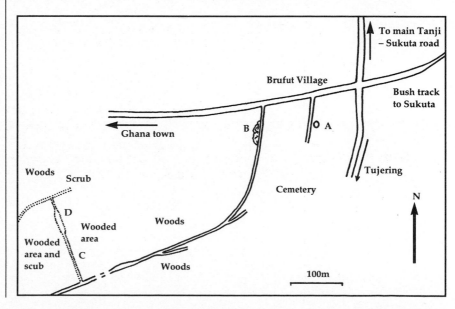

village you turn right from the laterite road, or carry straight on from the bush track. Look out for a narrow 'street' a little way down on the left, which leads to the village well - clearly visible from the main road, and take the next left turn down another narrow lane which leads past (and through) the village rubbish dump. Carry on down here, ignore any tracks which enter from the left, and keep bearing round to the right until you leave all signs of the village behind and enter a heavily wooded area.

The track has some very rough stretches to begin with, but is still negotiable with care and four-wheel drive and becomes more level eventually. It is very narrow, with one or two 'passing places' on the right hand side. Look out for a very narrow track leading off to the

White-faced
Scops Owl

right; this is a definite right turn, not a fork, and although it may be rather overgrown it is still possible to drive along it. It is best to get out here and walk, leaving the driver to follow later so as to cause least disturbance. After a short distance this track widens briefly, providing a convenient place to park without creating an obstruction.

Strategy   A late afternoon or early evening visit can be good for White-faced and African Scops Owls, and there is always a good variety of other species to be seen, including nightjars. Didric Cuckoos are breeding visitors and will not be found between November and June. Biting insects descend in hordes at dusk, and a good supply of repellant is essential.

Birds   The woods all around the track provide good birdwatching and it can be particularly rewarding to walk to the end where another track crosses, forming a T-junction. The scrub down to the right is a haunt of nightingales, while the surrounding trees may harbour cuckoos, woodpeckers, Lesser Honeyguide and Snowy-crowned Robin-Chat.

The trees around the 'parking area' are a good place to look for a roosting White-faced Scops Owl, while the distinctive piping call of the diminutive African Scops Owl is almost certain to be heard as dusk descends, although the bird itself may be difficult to spot among the foliage without a good torch.

Kotu Creek and Ponds

Kotu Creek is a small tidal creek, fringed by mangroves and bordered by the Fajara Golf Course to the north and rice fields and Oil Palms to the south. It is, for the most part, easily accessible and is a particularly good place to see a variety of herons, egrets and waders. The elusive African Finfoot has been recorded but the chances of seeing one are remote.

Kotu Ponds must be one of the few sewage treatment plants in the world to be marketed as a general tourist attraction. The birdlife is rich and varied; the ponds and their banks regularly harbour a good number of waders, particularly Black-winged Stilts, Wood and Marsh Sandpipers and Ruffs, as well as Long-tailed Cormorants and a resident flock of White-faced Whistling-Ducks. They are a regular haunt of Black Terns during the autumn months and White-winged Black Terns in spring. The surrounding vegetation, which includes trees, shrubs and a small cultivation, can contain parakeets, barbets and woodpeckers, while Little Swifts and Red-chested Swallows regularly hawk for insects over the water.

Location   Map Reference (Lower Bridge) - Sh10:152886. Kotu Creek lies immediately to the south of the main Kotu Beach tourist area and is crossed by two bridges, the Lower Bridge which carries the Kotu Beach

Key

Bungalow Beach Hotel	A
African Market	B
New Car Park	C
Bakotu Hotel	D
Kombo Beach Novotel	E
Kotu Strand Hotel	F
Lower Bridge	G
Beach Bar	H
Winner's Fast Food	I
Badala Park Hotel	J
Tam-Tam Night Club	K
Upper Bridge	L

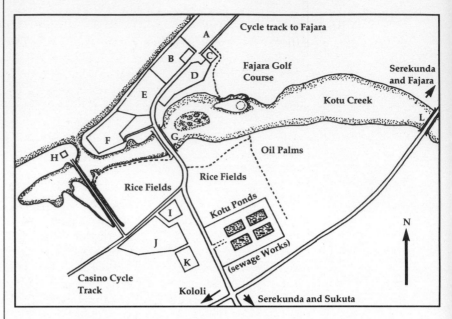

road itself, and the Upper Bridge about 1km upstream where the Kololi-Bakau road crosses. The Kotu Ponds are about 300m to the south of the Lower Bridge.

Bush taxis make regular trips from Serekunda to Kotu and it is best to get out at the Lower Bridge over the creek, just beyond the Badala Park Hotel.

From Cape Point it is probably as easy to get off a Serekunda-bound bush taxi as it turns onto Pipeline Road, and then walk back to the Fajara Club and across the Golf Course to the main Kotu Beach area - which is also the route for residents of the Fajara Hotel and Francisco's.

From the Palma Rima, the quickest way is along the cycle track at the back of the hotel which comes out by the side of the Badala Park.

From Kololi it is possible to walk along the beach in about forty minutes, or pick up a bush taxi for Serekunda, getting off at the Palma Rima and following the cycle track.

Strategy   The creek is best visited in the morning or evening, but the ponds can be good at almost any time. With the Casino Cycle Track and the Fajara Golf Course adjacent, the whole area is worth at least a couple of full days exploration. The Paradise Beach Bar, on the cycle track which runs along the shore between Kotu and Fajara, conveniently provides shelter from the heat at mid-day.

The charge for an admission ticket to the ponds is D5 and the 'ticket office' is a table beside an ancient cannon in the shade of a large tree, but the caretaker is often working in his vegetable patch and his attention may need to be attracted with a shout.

The best starting point for a walk is the Lower Bridge. In the early morning and late evening, this is the haunt of local birdwatching guides, licensed and otherwise, who often have a telescope set up and use the place as a base from which to tout for custom.

Birds   Scan the creek from both sides of the bridge. To the east, or upstream, there are wide mud-banks and areas of mangroves where a variety of waders is usually present, including Ringed, Grey and Spur-winged Plovers, Whimbrel, Common Sandpiper, Common Greenshank, Ruddy Turnstone and Little Stint. Wattled Plovers are frequently in evidence, while Senegal Thick-knees lurk in the large patch of mangroves near the middle of the area. Western Reef-Egrets and Great White Egrets are rarely absent, and the occasional Black Heron may also put in an appearance. Black-crowned Night-Herons roost in the tall trees to the left, which border the Bakotu Hotel gardens.

To the west, or downstream of the bridge, the muddy shores are narrower and fringed by mangroves on both sides. Malachite Kingfishers are regularly seen and there are usually a few waders. The telegraph wires are a regular perch for Wire-tailed and Red-chested Swallows and for Pied Kingfishers which are very common in the creek. Giant Kingfishers often fish from the wires, while Swallow-tailed and Blue-cheeked Bee-eaters also turn up from time to time, as does the occasional small raptor including Shikra, Black-shouldered Kite and Red-necked Falcon.

A narrow path leads from the bridge to the south bank of the downstream section and winds between the mangroves and what seem to have once been rice fields - now the haunt of Little Bee-eaters and Yellow-billed Shrikes. The path eventually meets with another which runs down to the shore from the Casino Cycle Track just outside the side entrance of the Badala Park Hotel. Getting to this means crossing a deepish ditch, which is sometimes wet or muddy depending on the state of the tide. Beyond this is a shallow tidal pool which is usually good for waders, including Black-tailed Godwit, and attracts a variety of terns during the spring and autumn.

Going towards the Badala Park, the path passes through a small area of rice fields, where Green-backed Herons may sometimes be seen. It then reaches the cycle track where the path to the left goes back to the road.

Turning back towards the bridge, watch for a path leading off right, down the steep bank and across the rice fields bordering the south side of the creek - these are home to Green-backed Herons and the occasional African Jacana. The path leads into an area of Oil Palms and down to the edge of the creek, where there are good spots for wader-watching.

The opposite bank borders, and is best reached from, the Fajara Golf Course. Walk on down the road to the bottom, and then along the cycle track that runs round to the left of the Bungalow Beach Hotel until the end of the wall which surrounds the car-park on the right. Turn right here onto the edge of the Golf Course and follow the wall of the car-park and then the boundary wall of the Bakotu gardens which soon turns sharp right. Here there is a wettish depression and beyond are the mud and mangroves which border the creek.

This is a good area for birdwatching, with waders, the chance of a Hamerkop, and a variety of smaller birds which inhabit the mangroves themselves and the scrub which lies between them and the Golf Course.

Giant Kingfisher

Yellow-crowned Gonolek, Beautiful Sunbird and Brown Babbler all regularly occur here, as well as warblers and the commoner estrildine weavers. It is possible to walk east along the edge of the creek, eventually ending up at a wooden bridge linking the Golf Course to one of its 'greens' which lies on a small island in the creek. The island is a convenient spot for wader-watching, with Wattled Plovers being especially common in this region, and the chance of an Osprey or Palm-nut Vulture on the far bank. A second bridge leads back to the shore, which can be followed as far as the Upper Bridge if time permits.

The entrance to the Kotu Ponds is opposite the 'Tam-Tam' night club and is clearly signposted. Just inside the entrance, on the left, are some scattered trees and a small, mixed cultivation. Snowy-crowned Robin-Chats have been seen here regularly, usually in the early morning, as well as Bearded Barbets and, in autumn, the occasional Woodland Kingfisher.

There are four pools altogether, of which the two on the right always seem to be full and those on the left vary, with at least one of them being either empty or nearly so.

Spur-winged Plovers and Black-winged Stilts are common on the first pool on the right, with a good variety of other waders, particularly Common, Marsh and Wood Sandpipers, Ruff and Common Redshank. Beyond this, the second pool has well-established marginal vegetation and is home to a flock of White-faced Whistling-Ducks, as well as Long-tailed Cormorants and occasional African Darters. The pools on the left are more variable but usually harbour an equally impressive variety of waders.

The tops of the banks between the pools are very overgrown just after the rains, and flocks of Northern Red Bishops and other weavers may be seen feeding on the grass heads in October and November. The dry vegetation is cleared and burned off in December and the weavers disappear, with Speckled Pigeons taking over on the open banks.

Other wildlife

The creek is home to Nile Monitors which can reach impressive size and are sometimes to be seen swimming in the shallows or crawling up the banks. They are extremely shy and can move remarkably fast when disturbed. The large Gambian Fruit Bat is often seen flying over at dusk, especially near the Kombo Beach Hotel.

Casino Cycle Track

The Casino in question is next door to Solomon's Bar, on the shore by the Palma Rima, but is no longer open. The so-called 'Casino scrub', which once lay behind it and was famous for the Bronze-wing Coursers which appeared there regularly in December, is sadly no more, having met its end at the hands of developers.

The track passes through a variety of habitats, including scrub, rice fields, open areas with scattered Oil Palms, and small groups of Acacias, and several of the commoner species of birds may be seen here.

Between the first part of the track and the shore lies an extensive area of Gingerbread Plum and low scrub, crossed by a network of wide, sandy tracks, which provides very rewarding birdwatching with kingfishers, barbets, woodpeckers and shrikes, the chance of a Oriole Warbler or a Pied-winged Swallow, and nightjars regularly seen at dusk.

At the Badala Park end, hidden away behind a row of young Rhun Palms, lies a small area of permanent standing water which attracts herons, egrets and a variety of waders.

Location

Map References: Palma Rima end - Sh10:142878; Kotu Creek end - Sh10:152885. The Casino Cycle Track runs for just over 1km, from the back of the Palma Rima Hotel to the Kotu Beach road, onto which it emerges by the side of the Badala Park Hotel, a short distance from the Lower Bridge over Kotu Creek.

From Kololi it is a fairly easy walk along the beach to the road which leads up to the Palma Rima from Solomon's Bar. For residents of the hotels at Kotu Beach, Fajara and elsewhere it is best to start at the Kotu Creek end, reaching this as described in the previous section.

Strategy

The ideal time for a visit is the early morning, when the birds in the coastal scrub are most active. This is the most interesting area and

Key

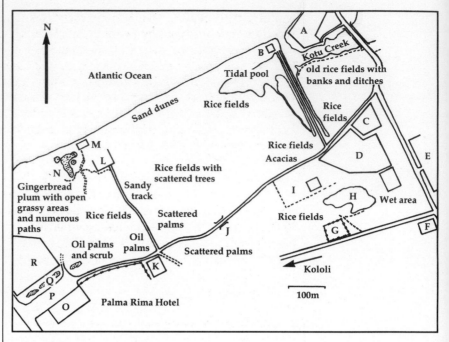

should be explored first, followed by the pools behind the Badala Park, although if the septic tanks at the back of the Palma Rima are in full flood, a quick check for Greater Painted-snipe should be made before there is too much disturbance. For nightjars, a visit at dusk will be required.

Birds

Starting at the Kotu end of the track, with the Badala Park Hotel on the left, there is low scrub with Acacias bordering rice fields to the right. The Acacias are usually good for sunbirds, estrildine weavers and visiting warblers, with Subalpine and Olivaceous Warblers being present throughout much of the dry season.

Just after the hotel, on the left, is a roughly quadrangular area of bare ground with a couple of baobabs, at the back of which is a row of young Rhun Palms. Behind this, and hidden by the palms, are rice fields and a swampy depression surrounding an area of standing water. This can be viewed from the bank behind the Rhun Palms, but it is better to cross the rice fields to higher ground which overlooks the pools from the opposite side (also reachable by a path from the Kotu-Kololi road).

The pools attract herons, egrets and waders, with Black and Squacco Herons being the most notable regulars.

Beyond this, the track passes beside rice fields and open uncultivated areas with a scattering of Oil Palms, where Green-backed Heron, Hamerkop, Lizard Buzzard, Double-spurred Francolin, Western Grey Plantain-eater, Green Wood-Hoopoe and Grey Woodpecker all commonly occur. Pearl-spotted Owlet is regularly recorded.

Shortly after the track crosses a culvert there is a large house on the left, and immediately before this a wide track leads down towards the sea, ending abruptly at a walled building site. A narrow path skirts round to the left and soon joins another track which runs alongside a wet depression fringed by Swamp Date Palms and a couple of fair-sized Acacias. A narrow path leads to the other side of this where an area of raised ground provides a convenient viewpoint.

The depression and the scrub around provide an interesting and productive birding area, with regular appearances of Bearded and Vieillot's Barbets, Yellow-fronted Tinkerbirds, Grey and Cardinal Woodpeckers, Nightingale, Singing Cisticola, Oriole Warbler, Orange-cheeked Waxbill and Black-necked Weaver.

The site borders an extensive area of Gingerbread Plum which lies between the cycle track and the sea and which is crossed by a network of paths which eventually lead to the back of the Palma Rima. Striped and Woodland Kingfishers, Little Bee-eater, Yellow-fronted Tinkerbird, Yellow-crowned Gonolek, Black-crowned Tchagra, Northern Crombec, Northern Black-Flycatcher and African Silverbill are all commonly seen here, while African Harrier-Hawk and Lizard Buzzard are the most likely raptors and dusk brings a good chance of both Long-tailed and Standard-winged Nightjars, especially near the dunes which fringe the shore.

The southern end of the cycle track leads into a wide, open, sandy area which lies behind the main bedroom block of the Palma Rima Hotel. On the other side of this area is an uncompleted apartment block, which seems likely to remain uncompleted for the foreseeable future, surrounded by a high perimeter wall. Between the wall and the open space there is a narrow belt of low scrub with several muddy depressions which periodically become swampy pools when the Palma Rima septic tanks overflow. This attracts waders which may include Wood, Green and Marsh Sandpipers as well as the occasional Greater Painted-snipe. Yellow Wagtails are also common visitors as well as Variable Sunbirds and the commoner estrildine weavers.

Oriole Warbler

Fajara Golf Course

The Fajara Golf Course occupies an elevated position with a steep bank on the northern side which looks out over a wide stretch of low scrub to the sea. Between the western end of this bank and the Bungalow Beach Hotel is an area of well-established Gingerbread Plums and taller trees which is particularly good for birds, including Pearl-spotted Owlet as well as a variety of bee-eaters, shrikes, babblers and sunbirds. Cattle Egrets and Black-headed Plovers are common on the fairways, and there are clumps and rows of a mixture of trees, including palms, which attract hornbills, Western Grey Plantain-eaters, Green Wood-Hoopoes and Purple Glossy Starlings, while Palm Swifts and Swallow-tailed Bee-eaters may often be seen circling above. There are good views of Kotu Creek from the south side, and the two areas can be combined into a single birdwatching trip.

Location   Map Reference (Kotu end) - Sh10:155889. The Fajara Golf Course lies between Kotu Creek to the south and Kotu Beach to the north, with the Fajara Hotel at the east end and the Bakotu at the west.

The Golf Course may be entered with equal ease from either the Fajara or the Kotu end, although the former is probably the more convenient starting point. The free-lance caddies will probably try to persuade you to play golf, but they are easily discouraged and will not continue to pester you. Beware of bogus 'birdwatching ticket' sellers (see p15).

Strategy   Early morning is the best time for a visit, and late afternoon can also be good. A visit here could follow a morning visit to either the creek or the cycle track and a late morning or early afternoon visit to the ponds.

Birds   When starting at the Kotu end it is best to walk first along the Kotu-Fajara cycle track which starts between the Bungalow Beach Hotel and the Sir William Restaurant. This runs along the shore and gives good views of the scrub below the northern edge of the Golf Course, where Black-crowned Tchagra and the occasional roller may sometimes be seen perched. Look out too for African Silverbills which frequent this area and may often be seen perched on the telegraph wires.

At its easternmost end the track turns sharply to the right, away from the sea, and reaches the bottom of a flight of steps alongside which a ramp has conveniently been provided for wheel-chairs. The high boundary fence of the Fajara Hotel is covered with purple Bougainvillea during the season and attracts many birds including the occasional Copper Sunbird.

Just beyond the top of the steps, a long row of Whistling Pines (Casuarina equisetifolia) stretches along the edge of a fairway to the right. One of these trees is the habitual roost of a White-faced Scops Owl and one of the caddies will point out the exact tree (for a small tip) - get here early in the morning to see it, as it gets mobbed by smaller birds and departs for more peaceful surroundings. This fairway is usually good for Black-headed Plovers, which are very common on the course.

Following the edge of the ridge along in the direction of the sixth tee provides good views over the scrub and Gingerbread Plums below,

Key

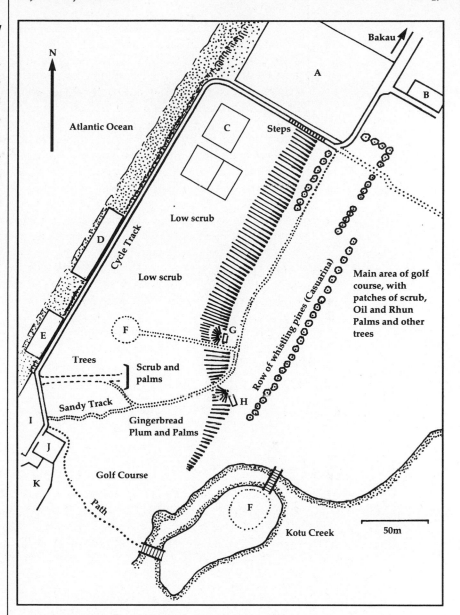

where babblers, warblers and sunbirds are usually common and there is a good chance of a Yellow-fronted Tinkerbird.

Just after the sixth tee, two well-defined sandy tracks lead down the hill, the first to the corresponding green (referred to as a 'brown' on this particular course) and the second to join the cycle track near the Bungalow Beach Hotel. Just after the second track is a steep knoll, topped with bushy Gingerbread Plums, at the far end of which lies the eighth tee, and this provides a good viewpoint from which to watch the birds in the vegetation below. An extended stop here, especially early in the morning, can be well worthwhile, for there are plenty of suitable perches and the area attracts a variety of species including Black-billed Wood Dove, Swallow-tailed and Little Bee-eaters, Bearded Barbet,

Yellow-throated Leaflove, Black-crowned Tchagra and both Brown and Blackcap Babblers.

Behind this bank, stretching away south-east, lies the main part of the Golf Course, with many areas of trees and shrubs. A return to the Kotu end of the cycle track may conveniently be made by following the edge of the creek in a westerly direction as far as an island on which lies one of the 'browns'. The island is reached by two bridges and from the second bridge a path leads towards the back of the Bakotu Hotel, through an area where Double-spurred Francolin are frequently heard and sometimes seen, eventually joining the cycle track by the side of the new car park.

Other wildlife   Green Vervet Monkeys and Gambian Sun Squirrels occur regularly, and there is the chance of a Ratel or Honey Badger, although this species is mainly nocturnal.

Black-headed plover

Atlantic Road and the MRC Grounds

The US Ambassador's Residence, the British High Commission and the President's residence are all located along the Fajara end of Atlantic Road, as are many smaller houses with well-tended gardens which attract a wide variety of birds, including Copper Sunbird and Chestnut-bellied Starling, while African Green Pigeon may also be found, especially in January when many of the trees are laden with fruit. Try and gain admission to the Medical Research Council grounds outside normal working hours - this is a peaceful spot with various birds including Shikra, which breed here regularly.

Location

Map Reference (MRC entrance) - Sh10:169901. Atlantic Road (also called Atlantic Avenue) runs for about 4km, from the Fajara Hotel at one end to the beginning of the Old Cape Road at the other. The MRC laboratories and clinic are about 1500m along the road from the Fajara end, and it is this section which provides the best birdwatching.

From hotels in the main Kotu Beach area it is an easy walk across the golf course or along the cycle track. From the Palma Rima and Kololi, the choice lies between paying for a tourist taxi to Fajara, or taking a 'tankatank' into Serekunda and another one out towards Bakau, getting out where the vehicle turns right off Pipeline Road and walking the short distance to Atlantic Road.

From the African Village and the hotels at Cape Point, it is a long walk down the entire length of the road, and it is better to catch a Serekunda-bound bush taxi and get off when it turns onto Pipeline Road.

Strategy

During January many of the trees lining the road are heavy with fruit and attract a good variety of species, but there are plenty of things to see during any month.

The MRC grounds are private property and there is no right of access, but genuine birdwatchers are rarely refused admission if a polite request is made. Please respect the privacy of the area and cause as little disturbance as possible. Several members of the MRC staff are or have been members of the Gambian Ornithological Society and take a special interest in the birdlife of the site.

The grounds are best visited outside normal working hours if possible, with Sunday being the ideal day and late afternoon the ideal time.

Birds

Starting at the Fajara Hotel end, the first main area of interest lies just after the junction with Pipeline Road where two driveways lead up the rising ground to the left.

The first runs between private houses and ends at the top of a sandy track which leads steeply down to a beach bar and the shore below; the second curves back on itself, again past a private residence, and ends in the yard of a compound. Between the two is an area of dense scrub and trees, while bordering them are the well-maintained gardens of the European-style residences.

There are usually plenty of birds to be seen flitting around here, with an abundance of the commoner sunbirds and estrildine weavers but also a chance of rollers, barbets, Leaf-love, Northern Puffback, Snowy-crowned Robin-Chat and Copper Sunbird, while on the opposite side of the road from the second driveway is an area of trees and cultivations where a similar variety of birds may be seen.

A little further down the road, the pumping station of the local water supply is situated on the right hand side with its water towers on a patch of open ground opposite. Just beyond the latter is an area of dense scrub and woodland, after which European-style houses and

Key

British High Commission	A
Medical Research Council Grounds	B
VSO Field Office	C
Private Houses	D
Pumping Station	E
Trees and Scrub	F
Petrol Station	G
US Ambassador's Residence	H
Fajara Hotel	I
Fajara Club	J

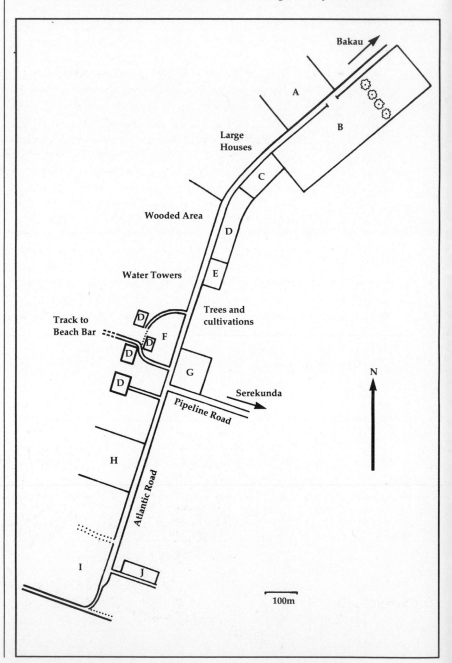

gardens occupy both sides of the road, providing a good-opportunity for a spot of 'garden-watching', with sunbirds and waxbills in good numbers as well as the chance of a Little Weaver or a Chestnut-bellied Starling, the latter a species which is virtually a speciality of the Fajara gardens. Tall trees bordering the avenue on the left hand side also attract good numbers of birds, especially when in fruit, with Bearded Barbet and Purple Glossy Starling especially common and the chance of something more exotic, such as a African Green Pigeon, lurking among the leaves.

The local VSO Field Office is situated on the right hand side, just after the road makes a turn to the right, and is followed, on the opposite side of the road, by the impressive home of the British High Commission, which faces the extensive grounds of the Medical Research Council.

The entrance to the MRC is guarded by tall, locked gates and a uniformed commissionaire in a booth on the left to whom a request for admission should be addressed. Just after the entrance there is a 'cross-roads' where a right turn takes the visitor along a tarmac roadway, bordered by mature trees, leading round the four sides of a square within which lie the laboratories and the clinic. The first two sides of the square run alongside the gardens of staff accommodation, and the central area, which is crossed by other driveways, contains a variety of trees and shrubs where there are usually plenty of birds, with a good chance of Shikra, Northern Puffback, and even a Lesser Honeyguide.

Returning to the main drive, turn right and walk to the opposite end from the entrance, where a left turn leads away from the buildings and into a more open area in the middle of which is the rugby football pitch. Around the edges of the sports field are plenty of wooded and shrubby areas, worth exploring, after which a left turn onto a driveway running along the Atlantic Road side of the field leads to the main entrance.

Chestnut-bellied
Starling

Old Cape Road and Camaloo Corner

The Old Cape Road passes through some excellent and popular bird-watching areas, including mangroves, cultivations and roadside scrub. During spring and autumn the area is particularly good for passage migrants, especially at the Cape Point end of the road.

At Camaloo Corner there are reed-beds, mangroves and mud-flats on one side of the road, holding good numbers of waders. On the other side, rice fields and marshes hold a variety of herons and egrets and a African Jacana is always a strong possibility. The scarce African Crake is regularly recorded from here but its skulking habits make it difficult to see.

Location

Map References: Old Cape Road - Sh10:189910 to 212892; Camaloo Corner - Sh10:205889. The Old Cape Road runs for some 3km, from the end of Atlantic Road in Bakau, to join the main Banjul-Serekunda highway about 2km from Oyster Creek. Camaloo (or 'Camalou') Corner lies at the junction of the Banjul - Serekunda highway with the main Bakau road, which meets Atlantic Road by the side of the CFAO supermarket.

Key

Amie's Beach and Sunwing Hotels   A

Yvonne Class Restaurant   B

CFAO Supermarket, Bakau   C

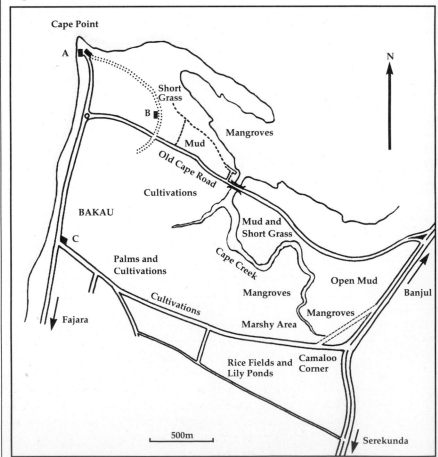

From the African Village Hotel and those at Cape Point, it is possible to cover all the best areas in a round trip of about 8km, by walking down the Old Cape Road, across to Camaloo Corner, and back up the Bakau road.

For anyone not relishing this amount of walking, as well as for visitors staying in Kotu or Kololi, it will be necessary to take a 'tankatank' to Serekunda and pick up one of the frequent Banjul-bound minibuses. Most drivers have never heard of 'Camaloo Corner' although they may possibly know it as 'Stink Corner', its rather unfair alternative name. It is easier to watch the road carefully after it leaves the industrial area of Kanifing and look out for a large green road-sign indicating the Bakau road to the left. Ask the driver to pull up at the junction, which is Camaloo Corner. The minibuses usually travel at a fair speed, so give plenty of warning.

Strategy

There is usually plenty to see at most times of the day, but because the road is the shortest route from Bakau to Banjul it can be very busy during the morning and evening 'rush hours', particularly with vehicles bearing the green 'CD' and red 'TA' (Technical Assistant) licence plates, which tend to tear along at ridiculous speeds considering the narrow and uneven nature of the road. Because of the direction of the sun, it is best to walk from Camaloo Corner towards Cape Point in the morning, and in the opposite direction in the afternoon.

Birds

Starting at Camaloo Corner and walking towards Bakau, the area on the left hand side consists of rice fields, between which lie swampy areas and pools of standing water with the water lily Nymphaea maculata much in evidence. This is a very good area for herons and egrets with Grey, Black-headed, Purple, Squacco and Green-backed Herons all likely as well as Great White and Cattle Egrets, Western Reef-Egret, and the chance of an Intermediate Egret, Sacred Ibis or African Spoonbill. Hamerkops occur regularly, and this is one of the best places in the region to see an African Jacana, a surprisingly elusive bird considering how common it is supposed to be.

A hundred yards or so up the road the swampy area ends, and this is a convenient place to turn back and concentrate on the mud-flats and mangroves on the opposite side. A number of waders can usually be seen, although not the variety found in some areas, with Black-tailed Godwit often present in large numbers. Osprey and Western Marsh Harrier can often be seen over the mangroves.

Just before the junction with the dual carriageway, a narrow section of the old road curves away to the left. Both sides of this are lined with shrubs, including low Acacias, and attract a variety of small passerines, including Orange-cheeked and Black-rumped Waxbills, especially in December when the Acacias are in flower. After a short distance the main highway is reached again, running alongside the mud-flats to the junction with the Old Cape Road. It is possible to take a short cut across the dried out mud once the dry season is well-established, but be sure to follow the trail of former footprints, for there are unexpected soft areas which can be extremely dangerous.

African Jacana

The flats between the Old Cape Road and the mangroves which border Cape Creek have areas of short grass which are the haunt of Crested Lark and the striking Yellow-throated Longclaw, while the creek itself may harbour Pink-backed Pelicans and sometimes a Purple Heron or an Osprey perched in one of the taller mangroves.

The road crosses the creek by a narrow bridge, on the far side of which a short track leads off to the right and down to the edge of the mud-flats and mangroves lying between here and the shore. A left turn at the bottom of this leads to a small cultivated area, bordered by mangroves, and an area of scrub, beyond which an open, grassy area can easily be crossed to rejoin the road. The whole area is rich in birdlife and a good place to see one of the less common passage visitors as well as species such as Winding Cisticola, Quailfinch and Black-headed Weaver.

Turn back onto the road in the direction of Bakau, where a dirt track soon winds off to the left through mixed cultivations. This eventually leads to a rubbish dump, and is worth a detour if time permits.

The main road soon enters Bakau, ending at a small traffic island where the road to Cape Point goes off to the right while Atlantic Road is to the left. On the left side of the latter is the local market, where the fruit and vegetables are of good quality and reasonably priced. A little further on is the CFAO supermarket and the Bakau bush taxi depot, from which vehicles return regularly to Serekunda.

The Bund Road

The 3km of the heavily pot-holed and generally disintegrating Bund Road pass a lagoon, tidal mud flats, mangroves, canals, ditches and some cultivations, and offer one of the most interesting, and perhaps least salubrious, birdwatching sites in the area. Greater Flamingo is the speciality here, with Yellow-billed Stork, Sacred Ibis and African Spoonbill all possible. There are always good numbers of waders on the mud flats while the telegraph wires along the roadside provide perches for bee-eaters, rollers and the occasional small raptor.

Location

Map References: West end - Sh10:290868; East end - Sh10:2718. The bund is actually a barrier with an associated pumping station which was constructed to reduce the risk of flooding in Banjul during the rainy season. The Bund Road, which runs along it, starts at the end of Cotton Street in the Banjul district of Half Die and ends at the main Banjul-Serekunda dual carriageway, a short distance from its junction with Independence Avenue and Box Bar Road.

The western end of the road is within walking distance of the Banjul Island hotels. From all others, it will be necessary to get a 'tankatank' into Serekunda and then catch a Banjul-bound minibus. Ask to be dropped at the junction of the Bund Road with the main dual carriageway, and start from here. On an early morning visit this means walking into the sun, which is not ideal but is compensated for by the fact that at the end of the road it is easy to walk through Banjul from Half Die to pick up a Serekunda-bound minibus at McCarthy Square. Working the road in the opposite direction will mean trying to stop a return minibus on the dual-carriageway, which is not an easy task.

Key

Atlantic Hotel — A

MacCarthy Square — B

Pumping Station — C

Lagoon (Crab Island Ponding Area) — D

Banjul – Barra Ferry Terminal — E

Wide Muddy Shore — F

Hulks — G

Cotton Street Bus Depot — H

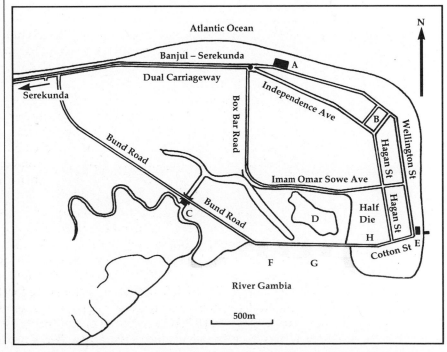

Strategy

During the working week, the road is a continuous stream of honking, rattling lorries, buses, taxis and private cars streaming out of the port area to reach the main highway and swerving disconcertingly from side to side in order to miss the worst of the pot holes. At weekends, in complete contrast, it is a veritable haven of tranquillity, making this the best time to birdwatch here.

Another factor to bear in mind is the state of the tide, as the mud flats at Half Die are quite extensive at low water, and a lot of the waders will be some distance from the road. Avoid mid-day when the sun is directly behind the estuary and mud flats, and the glare is unbearable.

It is probably best not to venture along the Half Die end of the road alone as a variety of strange characters may be encountered there, poking through the rubbish which lines the road.

Birds

Turning down the road from the main highway, there is at first little of interest, with flat, open areas, some under cultivation, stretching away on either side. These soon give way to dense mangroves on the right which persist as far as the Pumping Station where they suddenly open out into a creek, conveniently viewed from the bridge, where herons, egrets and waders may be seen on the mud banks on the estuary side, as well as the chance of a Yellow-billed Stork or a White-faced Whistling-Duck. Little Swifts regularly breed in the building itself. On the opposite side of the road, the creek is straight and narrow between mangroves on either side and runs eventually into a long stretch of water which crosses it at right angles, where Black Terns can usually be seen.

Greater Flamingo

Beyond the Pumping Station, the ditch is continuous on the left hand side of the road with muddy banks, which harbour the occasional wader or egret, lined by low bushes which attract a variety of smaller species. Telegraph wires run the entire length of the road, but seem particularly attractive to birds from this point onwards, with a good chance of Blue-cheeked and Swallow-tailed Bee-eaters as well as rollers and occasionally small raptors, including Black-shouldered Kite, and Grey and Common Kestrels.

Just over a quarter of a mile beyond the Pumping Station the road swings to the left to run alongside the shore of the Gambia River estuary. Here on the right there are wide mud flats at low tide which attract large numbers of waders including Curlew Sandpipers and occasionally Pied Avocets, while the hulks and assorted bits of boats which lie both on- and offshore provide perches for Pink-backed Pelicans, Long-tailed Cormorants and the occasional Great Cormorant as well as several species of terns. Great White Pelicans also occur here from time to time. The drainage ditch/sewer continues on the opposite side of the road, where the main attraction is the sizeable lagoon shown on some maps as the Crab Island ponding area.

This wide but shallow expanse of water is fringed by tall mangroves, which make visibility a problem in places. It is possible to reach the other side, where drier areas provide a chance of Kittlitz's Plover, by braving the backstreets and infants of Banjul, but it is probably best not to go without a local guide. Adequate views of most of the lagoon can be obtained from a number of points along the road. Apart from the usual herons and egrets, there is a good chance of seeing Greater Flamingo and a possibility of Yellow-billed Stork, Sacred Ibis and African Spoonbill, while a variety of waders, including Black-winged Stilt and Marsh Sandpiper, frequent the shallow margins and Greater Painted-snipe are also regularly recorded from the area.

Beyond the lagoon, the Bund Road runs into Cotton Street in the port area of Half Die. At the beginning of Cotton Street is the main bus terminal for the south bank of the river, from which the buses for Soma and Basse depart. Just beyond this, off to the left, is Hagan Street, which runs directly to join Independence Avenue on the corner of McCarthy Square, from where the minibuses back to Serekunda depart.

Lamin Fields

This area's sole claim to ornithological fame lies with the Temminck's Coursers which appear here regularly after the water melon fields have been nearly cleared in December.

Location

Map Reference - Sh10:215790. Lamin fields lie just off the main road beyond the Girls' Technical School, at the southern end of Lamin village. Bush taxis bound for Brikama from Serekunda pass the end of the track which leads to the fields, from where it is a relatively short walk to the fields themselves. Ask to be dropped at the Technical School. Bush taxis

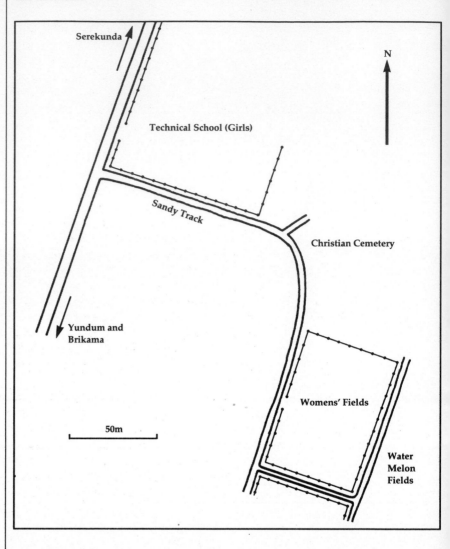

bound for Serekunda stop regularly in Lamin to drop off passengers, so there should be no problem getting one for the journey back.

Strategy　It is necessary to get to the site as early in the morning as possible, before there are too many workers around.

Apart from the Temminck's Coursers, the area is not especially rich in birdlife and is only worth a brief visit on the way to Yundum or Pirang.

Birds　A broad, sandy track leads from the main road along the side of the southern boundary fence of the school, eventually curving sharply to the right at the corner of the fence where another track goes off to the left. On the left hand side here is a small Christian cemetery where Black-headed Plovers are often seen, and just beyond this lie the women's fields, surrounded by a high fence.

Follow the track along past the gated entrance to these fields, to the point where the boundary fence turns left at the far end. Between this and the boundary of the next cultivation there is a narrow and overgrown-looking path, which leads up a slight rise to the extensive water melon fields which lie behind the fenced off areas. The cultivations on either side are worth a brief examination for estrildine weavers, bishops and warblers which frequent the variety of bushes and shrubs, while the occasional shrike or roller may be found perched on the fence.

The fields themselves are flat and open, with a few scattered trees, and are easily viewed from a track which runs along the edge.

Yundum

The most interesting site at Yundum lies behind and to the south of the Agricultural College complex, where a network of tracks criss-crosses a variety of habitats. Much of the area consists of low scrub with scattered Acacias, where a variety of small passerines may be seen, but there is also a small Acacia wood, which can be particularly good in December when the trees are in flower. Cultivations which attract Temminck's Coursers after the ground has been cleared, and a variety of trees and shrubs, some in the grounds of the college itself, are visible from a track running alongside. The area is also quite good for raptors.

Location   Map Reference (Turn off for Old Yundum) - Sh10:193765. The area lies to the south and west of the Yundum Agricultural College which is on the main Banjul Basse road, just over 1km beyond the turn off to the airport. Minibuses running to Brikama from Serekunda pass the site fairly frequently. Watch out for the entrance to the International Airport on the left and then the army barracks followed by the Agricultural College complex on the right. A short distance beyond the college grounds and on the opposite side of the road is the end of the main runway. Ask to be dropped off here. The circular walk from here arrives back on the main road by the side of the college from where a bush taxi can be stopped for the return trip.

Strategy   Early morning or late afternoon is the best time for a visit. Avoid weekends, or early afternoon when schoolchildren are finishing, otherwise a small but persistent party of hangers-on will appear. If time permits, explore the area around the old airport road which leads off from the main road immediately before the modern airport entrance and on the same side. This is now very much overgrown and access is not as easy as it used to be at one time. Once the crops have been cleared in December, the fields bordering the main road in the area of the police check point are a good place to look for Temminck's Coursers.

Birds   Close to the end of the main runway of Yundum International Airport, and on the opposite side of the road, two sandy bush tracks, a few yards apart, head off in the direction of Old Yundum.

Although the two tracks merge after a few yards, then divide and finally merge again, the northernmost of the two runs along the edge of

Key

Agricultural College Grounds with Citrus Orchard	A
Small Cultivations	B
Inhabited Buildings	C
Deserted Buildings (Military appearance)	D
Laterite Tracks	E
Sandy Bush Tracks (Main ones only are shown)	F
Runway Light in Small Enclosure	G
Main Runway	H

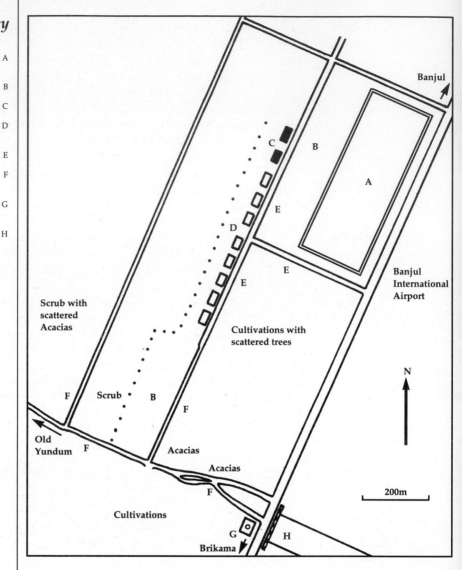

a small Acacia wood with cultivated areas between the trees and offers the best birdwatching opportunities. Warblers, sunbirds and estrildine weavers are usually in abundance, especially when the trees are in flower, while Levaillant's Cuckoo is a strong possibility in the early part of the season and Great Spotted Cuckoo has also been seen here.

Just after the wood, a well-defined track leads off to the right, narrow and sandy at first but soon widening into a good laterite surface. Ignore this for now and head straight on along the rather rutted bush track in the direction of Old Yundum Village.

This track passes between cultivated areas and low scrub where Northern Red and Black-winged Red Bishops abound, although they are nearly indistinguishable out of breeding plumage. Yellow-shouldered Widowbird may also be found and there is a chance of seeing a Chestnut-crowned Sparrow-Weaver here.

On the right hand side of the track, a number of side tracks lead off into the low Acacia scrub, and a short detour along one of these may produce a Siffling or a Singing Cisticola or even a White-fronted Black Chat, while Black-crowned Tchagras are common in this area. Although about 2km to Old Yundum itself, it is worth walking the whole route since this is an area where all sorts of interesting things may be seen, with Long-crested Eagle appearing regularly as well as several other raptors and the chance of a Pale Flycatcher or a Yellow Penduline Tit in the more open wooded areas.

Returning towards the main road, a left turn onto the track by the Acacia wood runs at first between cultivations, where Temminck's Coursers are regularly reported after the ground has been cleared in December and where Wattled and Black-headed Plovers also occur. While cultivations with scattered trees and shrubs continue on the right hand side of the track, the left soon becomes occupied by a row of deserted buildings which have a distinctly military air about them. There are plenty of trees and shrubs around, attracting a variety of species such as Northern Puffbacks and Yellow-fronted Canaries.

Where the cultivations on the right hand side end, a well-surfaced track leads off to the right, providing a convenient way of returning to the main road. It is bordered on the left by a narrow band of trees and shrubs which separate it from the boundary fence of the Agricultural College grounds. On the right a strip of low scrub with occasional trees divides it from the cultivations. There is a particularly fine Red Silk-cotton a little way down which attracts a host of birds, including Scarlet-chested Sunbirds, when the flowers open in January.

Long-crested Eagle

Kabafita Forest Park

Like the majority of the so-called 'Forest Parks' in The Gambia, Kabafita is not officially open to the public, but it is crossed by a number of tracks which make access relatively simple. Like the Nyambai Forest Park on the opposite side of the road, most of the woodland consists of monotonous, and ornithologically unproductive, plantations of the introduced, fast growing Gumbar, but there are some smaller areas of relatively undisturbed natural savannah woodland which can turn up interesting species from time to time.

Key

Grassy Triangle at Turn off	A
First obvious Cross Path	B
Wooded Area with Scrub and LargerTrees	C
About 1500m from Main Road to this point	D
Obvious Isolated Rows of Gumbar	

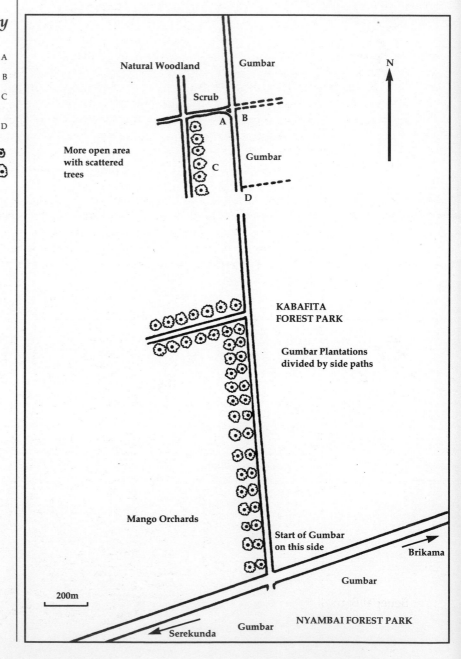

Location

Map Reference (Turn off) - Sh10:197711. Kabafita Forest Park is situated on the east side of the main Banjul-Basse road about 4km before the town of Brikama. Minibus bush taxis pass the forest on the way to Brikama from Serekunda. From the drop-off point it is then about 1500m up the track to the best birdwatching area. Returning the same way should not be difficult, since it should be possible to stop a returning vehicle on the opposite side of the road. However, it may be necessary to walk to Brikama to find one with a spare seat, since there are few places between the town and the forest where passengers are likely to disembark.

Hiring a taxi for the trip is quite expensive but eliminates any risk of having an additional long walk along the main road. To drive into the forest to the best area will require a four-wheel drive vehicle, since the track is deeply rutted and uneven in places.

Strategy

Early morning or late afternoon visits are best, but avoid any overcast days when, as at Bijilo, the forest may seem completely devoid of birds. With independent transport, it is worth making a brief stop on the way at Yundum football pitch, on the left hand side of the road in the modern village of Yundum, which lies a few hundred metres beyond the end of the airport's main runway, as Chestnut-bellied Starlings are often seen here.

The track into Kabafita Forest lies on the left hand side of the road (going towards Brikama), where a mango orchard ends and the main Kabafita Gumbar begins. At first sight the track appears to be double, with a row of Gumbar running down the middle, but if driving in, keep to the right-hand side between the central row of trees and the main plantation.

Senegal Batis

After a short distance, a well-defined side track leads off to the left, between two rows of Gumbar, while the main track heads straight on along the edge of the wood. Narrow tracks lead off to the right at intervals, dividing the plantation into blocks, but about 1500m from the main road there is a well-defined cross track, with the left hand branch enclosing a small triangular area of ground. There is just enough room to park a vehicle on the right without obstructing the track and also sufficient room to turn round.

Birds

The area of natural woodland to the north of the track holds a good number of species and is worth an extended exploration, with the possibility of African Pied Hornbill, Brown-backed Woodpecker, White-breasted Cuckoo-Shrike and African Yellow White-eye.

The track to the left of the parking spot leads through dense scrub and large trees, where there are usually warblers, sunbirds and estrildine weavers, with a good chance of a Greater Honeyguide. It shortly meets another cross track. The left hand branch has a single line of widely spaced Gumbar separating it from a heavily wooded area with a tangled scrub understorey, and this a particularly productive region, with species such as Didric Cuckoo, Northern Puffback, Sulphur-breasted Bushshrike, Senegal Batis and African Paradise-Flycatcher being seen on a regular basis. The plantations and cultivations on the other side of this track should not be ignored, as they also attract a variety of birds including Scarlet-chested Sunbird.

Pirang

The main claim to fame of Pirang must be its flock of Black Crowned-Cranes which regularly roost in the mangroves beyond the village. Brown-necked Parrots nest here, about the nearest point to the coast where they are likely to be seen. The area also attracts a good variety of raptors.

The Scan-Gambia shrimp-farm occupies a large area to the north-east of the village. Up until now it has attracted good numbers of birds, but the company closed down in late 1992 and the future of the site is uncertain. Apart from the extensive area of mangroves bordering the creek, there are rice fields, Oil Palms, and open areas with scattered Acacias and other trees. The variety of habitats makes an extended visit worthwhile.

Location

Map Reference (Bridge) - Sh10:346672. Pirang village is situated just off the Banjul-Basse road, some 11km beyond Brikama and about 6km from Madina Ba and the turn-off for the Gambia-Senegal border and the water-holes of Seleti. The village itself is large and rather straggling, and contains what is supposed to be the largest Kapok tree in The Gambia.

Buses running from Banjul to Soma pass the turn-off for the village, from where it is a walk of about 1800m to the creek, which eventually

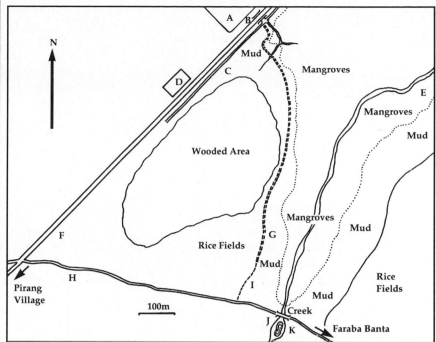

runs into the Pirang Bolon. If going by bush taxi from Serekunda either look for one departing for Soma from the main depot next to the market, or take a minibus to Brikama and change there, again disembarking on the main road and walking.

To really make the most of the area it is far better to hire a taxi for the day, or better still a four-wheel drive vehicle which will give the opportunity to combine a visit here with a drive across the bush from Faraba Banta to Jiboroh Kuta and possibly a visit to the Seleti water-holes as well.

Strategy   Avoid visiting the area on a Friday (when the Moslem school is closed) or a weekend (when all the schools are closed) as the village seems to have an unusually large, and persistent, juvenile population.

It is about an hour's drive to Pirang from the Bakau/Fajara area and it is best to arrive by dawn for a good chance of seeing the cranes.

After 5.7km from the police check-point at Madina Ba, look out for a wide sandy track leading off to the left by the side of a large 'Scan-Gambia' sign (which should still be there even if the company is not). The track leads through the village and eventually, about one kilometre from the main road, arrives at a fork, where a narrow, very straight laterite road leads left to the defunct shrimp-farm while a sandy bush-track heads right towards Faraba Banta. The creek is about 700m down here and is crossed by a dilapidated concrete bridge. If in a hired vehicle, it is best to park off the track just before the bridge, and explore on foot.

Birds   Shortly before the bridge, a footpath leads off to the left between the muddy edge of the creek and the rice fields. This is rather indistinct at first, but soon becomes clearly defined and runs along the top of a low

Black
Crowned- Crane

bank. On the left of the path, the initially open rice fields quickly give
way to more extensively wooded areas, with mixed Oil Palms and
other trees, where parrots and parakeets are frequently to be seen,
along with all four resident species of roller and the chance of an
African Pied Hornbill.

   The path follows the edge of the trees round to the left, until the
bridge is out of sight and the shrimp ponds come into view in the
middle distance. There may be tall elephant grass closely enveloping
the path for a few yards, but this soon ends, giving good views over the
mangroves to the right. Little Bee-eaters are common here, as well as
Yellow-crowned Bishops, although the latter are not easy to spot
among the dense foliage once the males lose their distinctive breeding
plumage. Far away across the mangroves, a tall, dead tree stands out
against the skyline. This is the favourite perch of a group of Black

Crowned-Cranes, but in the early morning it is directly into the sun, and a much better, and closer, view can be obtained from the other side of the creek.

As the edge of the trees curves away to the left, the raised path continues straight on past an open muddy area, which may have some standing water and which attracts small numbers of waders. Ahead is the laterite road which runs on an embankment alongside the shrimp ponds, and slightly to the left is a group of buildings which were the Scan-Gambia offices. This area is especially good for hirundines, with Red-rumped, Pied-winged, Red-chested, Mosque and Wire-tailed Swallows all being regular visitors, while Gull-billed Terns and others often circle over the shrimp-ponds.

On the skyline beyond the shrimp farm, tall trees on the edge of a wooded area provide perches for raptors such as Osprey and Palm-nut Vulture while flocks of Pink-backed Pelicans often circle in the sky overhead.

A sharp right turn along the edge of the mangroves brings the path gradually closer to the laterite road, which is eventually separated from it by a narrow channel where Malachite Kingfishers sometimes fish. It is possible to cross onto the laterite at this point in order to reach the shrimp ponds, where Black Crowned-Cranes often occur on the banks as well as Chestnut-backed Sparrow-Larks, Crested Larks and White and Yellow Wagtails. From here it is best to return by the same route to the bridge.

The creek to the right of the bridge, or 'upstream', has a small area of low mangroves among which a few Senegal Thick-knees are usually to be seen. The surrounding scrub harbours Four-banded Sandgrouse. Just over the bridge it is possible, once the dry season is well-established, to drive off the road onto firm mud and follow this around the edge of the bolon for some distance until the view opens out, with the mangroves ahead and to the left, and a wide area of cultivations stretching to the edge of the trees on the right. Park here and walk along the field edge.

The prominent tree which often provides a good view of Black Crowned-Cranes should be clearly visible from here, and usually provides a roost for Black-headed Herons as well, while the fields to the right may reveal Chestnut-backed Sparrow-Lark, Crested Lark, Plain-backed Pipit, Cut-throat and Quailfinch, especially after the crops have been cleared in December. The surrounding trees are favourite perches for several species of raptor including Short-toed Eagle, Dark Chanting Goshawk and Long-crested Eagle.

Beyond the creek, the bush track winds its way towards Faraba Banta, and if in independent transport, this is a good way to get back to the main road. After a short stretch of trees and rice fields there is a more open area to the right where cattle and donkeys often graze. Beyond this, the school football pitch will be seen on the left, and an enclosed area (the school grounds) on the right. Just before reaching these, there is an African Locust Bean just off the track, beneath which it is possible to park in the shade and obtain a fine view of the surrounding area.

White-billed Buffalo Weavers nest near-by, and a large and noisy flock is often in the vicinity, while a scan of the cattle and the donkeys usually reveals a few Yellow-billed Oxpeckers. Hamerkops are often here, and it is also a good place to spot a Dark Chanting Goshawk perched in a tree, especially on the school-side of the road.

The track leads on past the school and into Faraba Banta, where a right turn at the cross roads in the centre of the village eventually leads back to the main road.

Bush Track from Faraba Banta to Jiboroh Kuta

The 10km long, sandy bush track, which connects the Banjul-Basse road at Faraba Banta with the Madina Ba-Seleti road at Jiboroh Kuta, runs through a wide variety of habitats including cultivations, parkland savannah, and the untouched savannah woodland of the Finto Manereg Forest Park. The track is used only by the occasional ox-cart and passes through an impressively peaceful and unspoiled area of the countryside. It is also a marvellous place to see raptors, with a reasonable chance of a Bateleur or a Martial Eagle, and perhaps a Brown Snake Eagle or Gabar Goshawk. Like Kabafita, the Finto Manereg Park is not officially open to the public, but the bush track runs along the very edge for some distance and a number of side tracks provide access and the chance of seeing a few scarce residents such as White-winged Black Tit, African Yellow White-Eye and Black-faced Firefinch.

Location

Map References: Faraba Banta end - Sh22:346649; Jiboroh Kuta end - Sh22:287584. The northern end of the track leaves the Basse road some 15km east of Brikama, while the southern end connects with the Seleti road just over 10km south of Madina Ba and about 2.5km from the Gambian border post.

Because of the length of the track and its isolated nature, it is not feasible to visit using public transport, so it will be necessary to hire a vehicle. Since there are several stretches where the sand becomes very soft as the dry season progresses, a four-wheel drive vehicle is essential. Morfa Sunneh, who runs a four-wheel drive Suzuki, knows the route and can be contacted in The Gambia by phoning 94326.

If coming from Pirang along the bush track through Faraba Banta, it is a simple matter of crossing straight over when the main Banjul-Basse highway is reached. If coming directly along the main road, then look out for the Scan-Gambia sign and the turn-off for Pirang on the left. About 3.5km after this, the Jiboroh Kuta track goes off to the right immediately opposite the track to Faraba Banta. The local pottery seems to be on the left here, and there are usually several large water pots by the roadside, providing a convenient land mark.

Key

Small Enclosure	A
Northern Boundary Fence of Forest Park	B
Small Shelter	C
Small Track into Forest	D
Small Wooded Areas	E
Southern Boundary Fence of Forest Park	F
Short Detour around Fallen Tree (1993)	G

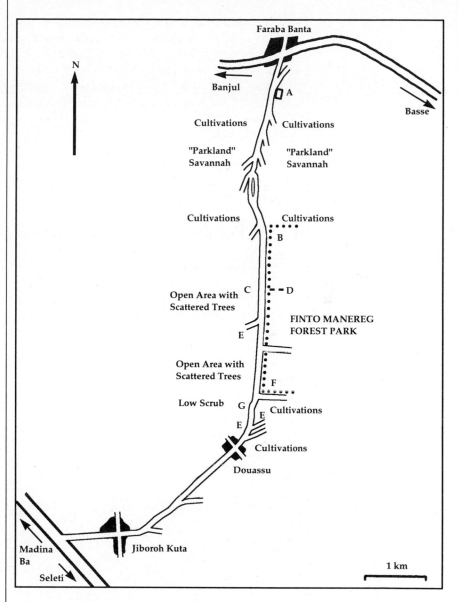

Strategy   Grasshopper Buzzard and Black-shouldered Kite are unlikely to be seen before November when there can be a sudden influx, and the former will not be around after February, while October and November seem to be the best months for Bateleur. However, there is a good variety of raptors around throughout the year and because of the open nature of the area and the amount of visible sky they are easily observed in flight. For this reason, the mid-day period, when the thermals are well-established, is much the best time for a visit. For the same reason dull, cloudy days should be avoided if possible.

There are several good stopping points along the track, with plenty of shading trees and the entire trip can take two or three hours to complete. This excursion can conveniently be combined with a morning visit to Pirang.

Birds

After passing a couple of compounds on the right hand side, the track leads into an area of cultivations with scattered trees which before long gives way to open parkland savannah. Here the landscape is dotted with Acacias, Red Silk-cottons and African Locust Beans which attract several species of birds, including a variety of warblers, waxbills and glossy starlings, while Abyssinian, Blue-bellied and Rufous-crowned Rollers are frequently encountered. Namaqua Doves are

Lizard Buzzard

common all along the track, especially from November onwards.

The whole stretch between here and the beginning of the forest park is especially good for raptors, and there are shady trees at convenient intervals along the track where stops can be made. The flat and relatively open nature of the countryside means that visibility is extremely good, and there are plenty of perches available for Dark Chanting Goshawks, Lizard Buzzards and Grey Kestrels which are all relatively common along here together with good numbers of Grasshopper Buzzards from November onwards.

As the day advances and the thermals become established, the larger birds of prey begin to put in an appearance, with a fair chance of seeing a Bateleur cruising low overhead, or an impressive and unmistakeable Martial Eagle, while a Long-crested Eagle is a frequent sight, the big white patches on its black wings showing up from a mile away. Western Marsh Harrier and Tawny Eagle are not uncommon, and Brown Snake Eagle, Gabar Goshawk and Red-necked Buzzard have all been seen here. After several days birdwatching on the coast, it is easy to overlook vultures, assuming almost automatically that they will be Hoodeds, but here it is worth checking every one, for there is always the chance it might turn out to be a White-backed, since individuals frequently circle in company with the commoner species. After about 3km the boundary of the forest park appears on the left. Much of this is the inevitable Gumbar, but there is a fringe of untouched savannah woodland, varying considerably in width, between this and the track. There is a low surrounding fence, but large stretches of this are broken down and, in any event, there are plenty of paths and tracks which lead off through the trees. Not far along here the track widens suddenly, becoming double for a short stretch, and a little further on is a low, rough shelter, where children who tend the local cattle are often to be seen. Immediately opposite this a narrow track leads off through the trees. It does not cover any great distance before reaching the Gumbar, but the trees and scrub to either side can be productive, with Northern Puffback, African Golden Oriole, barbets and warblers all regularly seen.

The main bush track continues along the edge of the forest with a more open area to the right before briefly being surrounded by savannah woodland on both sides. Just beyond this a well-defined track leads off to the left, and here it is a good idea to get out and walk, sending the vehicle on ahead to wait just beyond the forest boundary, which is situated at the next obvious side track and is delineated by the remains of a fence.

White-winged Black Tits occur in the area and parties may sometimes be seen moving restlessly through the trees. The many different species of trees which grow in this area come into flower and fruit at different times, and there are always some to attract feeding birds. The fruit-bearing trees tend to be worked fairly methodically by barbets and others, which then move on to the next supply, so it can be a matter of luck whether or not they are feeding close to the road. As everywhere, flowering Acacias attract warblers, sunbirds and waxbills, and ten or more species may sometimes be found in a single tree, including, perhaps, the elusive African Yellow White-eye or a Black-faced Firefinch.

The stretch of the main track which lies between this side turning and the next has dense savannah woodland on the one side and on the other a much more open area with patches of low scrub and scattered large trees. Violet Turaco, Eurasian Hoopoe and Abyssinian Ground Hornbill are three less common species which may be encountered along here, as well as Striped Kingfisher, Vieillot's Barbet, White-crested Helmet-Shrike, Bush Petronia and African Golden Oriole.

At the southern edge of the forest park, the next side track runs along beside the boundary fence and soon opens out on the right hand side where there is a large cultivation surrounded by trees, with some good perches for rollers and small raptors, while bee-eaters, warblers and sunbirds are usually in evidence and Levaillant's Cuckoo is regularly present early in the dry season.

Just beyond this turn-off, the main track is interrupted in its course by a fallen tree (although this is slowly being reduced to fire-wood and may not be there much longer) and makes a brief detour before passing briefly through dense savannah woodland on both sides, a favourite haunt of parrots and parakeets, to emerge into the cultivations which precede the small village of Douassu. There is an area of bare ground here with wooden stakes where cattle are tethered overnight. The hordes of flies which at once descend make leaving the area rapidly an attractive proposition, but there are often rollers and other insectivores around as well as the occasional small raptor.

After Douassu the track widens and the surface improves, passing through rather uninteresting cultivated areas before reaching the village of Jiboroh Kuta and, about 1km further on, the remarkably well-surfaced main road where a left turn leads to the Senegal border and the Seleti water-holes, while a right returns to the Banjul-Basse road at Madina Ba.

It is worth watching the sides of this road in both directions, since raptors often perch in the bordering trees, and there is a chance of seeing Abyssinian Ground Hornbills once the cultivations have been cleared in December.

Other wildlife

Green Vervets and Western Red Colobus Monkeys are common in the wooded areas while Patas may often be seen on the roadside. Long stretches of the bush track are bordered by fragrant herbs, some of which are used locally for making herbal tea and which are often alive with butterflies, especially in October and early November when African Monarchs, Black and Yellow Pansies, Citrus Swallowtails and a myriad of other species may be seen.

The Seleti Water-holes

The water-holes at Seleti lie just on the Senegalese side of the
Gambia-Senegal border. The largest, and the closest to the frontier,
holds water well into the dry season and sometimes right through the
year, but the two smaller ones which lie a little further to the south are
practically dry by late December. The water-holes attract a variety of
birds, mostly relatively common ones including large numbers of Red-
eyed, Laughing and Vinaceous Doves but also the occasional scarcer
species, and, their main attraction, Four-banded Sandgrouse which
regularly arrive to drink at dusk.

Apart from the water-holes, there is an excellent area of natural
savannah woodland which is easily accessible along a well-defined
track and which is home to a number of interesting species.

Location

Map Reference (Border Post) - Sh22:290561. The main water hole lies
a few hundred metres south of the Gambian border post on the Madina
Ba-Seleti road. It is on the right hand side immediately after the
'Senegal' border sign. The two other water holes are a little further
down the road, both on the left hand side.

Key

Border signs	A
Large Water-Hole	B
Small Water-Hole	C
Well-defined Track into Woodland	D
Medium-Sized Water-Hole	E

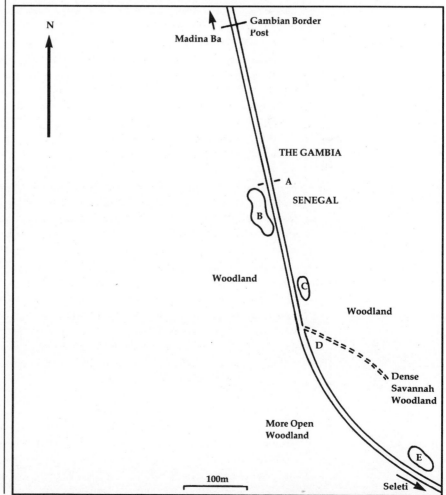

As with Pirang, the easiest way of visiting the site is to hire a taxi or a four-wheel drive vehicle. It is theoretically possible to visit the site using public transport, as the Banjul-Basse buses pass through Madina Ba and bush taxis run down to the border from Brikama. However, getting there and back in a day with reasonable time left for birdwatching is extremely difficult, especially if the main object is to see the sandgrouse, which only appear at dusk, as there is nowhere in the vicinity where overnight accommodation can be obtained.

Strategy

To see the sandgrouse it is essential to be there at dusk, when they fly in to drink at the water-holes. If a four-wheel drive vehicle is hired for the day, it is possible to combine a morning visit to Pirang with 'raptor time' in the bush and a late afternoon/evening visit to have a walk around the area and catch the sandgrouse as they arrive. However, a proper exploration of the surrounding forest will require at least half a day to itself.

Up until December 1992 there were absolutely no problems with crossing the border to birdwatch at the water-holes. The Senegalese border post is in the next village, some distance down the road, and the Gambian officials were quite happy to let people through on their side with no need for passport or visa. Then trouble with the separatist movement in Casamance (the southern region of Senegal) and the arrival of new staff at the border post resulted in requests for passports to be shown. The situation may revert to its earlier informality, but take your passport along to be on the safe side. A Senegalese visa is definitely not required.

Birds

The first, and largest, of the water-holes lies to the right of the road immediately after the 'GAMBIA' and 'SENEGAL' border signs, a few hundred metres down from the Gambian border post. There are suitable places for observation at both ends, and there is little to choose between them. The northern end is where the cattle come to drink, and the southern where the village children come to swim, so both are subject to periodic disturbance but the cattle are the more regular visitors.

There are usually a few Wattled Plovers around the banks and Purple Glossy Starlings often appear in large numbers. Other visitors may include rollers and the occasional raptor, with the chance of a Tawny Eagle, while Double-spurred Francolins are regularly heard and sometimes seen. An overhanging bush at the north end is often used by kingfishers, with both Malachite and Blue-breasted being possible. The swifts circling overhead will probably turn out to be Little but may include an Mottled Spinetail.

The next water-hole down lies on the opposite side of the road and is the smallest of the three. For some reason it seems unattractive to most birds but is the one which the sandgrouse visit most frequently and in the largest numbers. They do not start coming in until it is almost dark, and the best way to see them is to park on the roadside, which overlooks the water, and wait in the vehicle, getting the driver to switch on the headlights at the critical moment. Sudden illumination does not

Four-banded
Sandgrouse

seem to disturb the birds and extremely good views may be obtained.

Just down the road from this water-hole, and on the same side, a well-defined track leads into a large area of undisturbed savannah woodland, which is worth an extended visit if time allows. Among the raptors, Dark Chanting Goshawk and Grey Kestrel occur regularly, with a chance of something more unusual such as an African Hawk Eagle. The area has several species of woodpecker, including the scarce Brown-backed, as well as African Green Pigeon, Violet Turaco, White-breasted Cuckoo-shrike, Lead-coloured Flycatcher and White-winged Black Tit, while the elusive Stone Partridge inhabits the surrounding bush.

The third water-hole is also on the left hand side of the road and a little further down. The intervening roadside has plenty of bare trees for perching birds and these are often occupied by rollers or small raptors. Although the majority of visitors to this particular pool tend to be doves, there are more interesting arrivals from time to time with the occasional Broad-billed Roller, African Paradise-Flycatcher or Pin-tailed Whydah flashing down briefly to drink.

Other wildlife   Green Vervet and Western Red Colobus Monkeys are common around the pools. Western Baboons used to be seen regularly but persecution by local farmers has caused them to be much more circumspect, although they still appear occasionally.

Tendaba

The village of Tendaba lies just about on the border between the Lower and Middle River regions. The surrounding area offers the opportunity to see several species not found nearer to the coast. The rice fields bordering the Gambia River to the west of the village usually have a good variety of water birds, particularly early in the season, while the area around the airfield holds Bruce's Green Pigeon, Abyssinian Ground Hornbill and Yellow-crowned Bishop, with African Fish Eagle possible anywhere in the area. The creeks on the opposite bank of the river are particularly interesting - Goliath Heron is common and there are regular sightings of Blue Flycatcher, Mouse-brown Sunbird and the skulking and elusive African Finfoot, while African Swallow-tailed Kite may sometimes be seen.

Location

Map Reference (Tendaba Village) - Sh13:124858. The village is situated on the south bank of the River Gambia about 135km from Serekunda. It lies at the end of a laterite road about 6km from the junction with the main highway at Kwinella.

Some tour operators offer two day trips to Tendaba, staying overnight at the Bush Camp (see below). If this is all the time you want to spend there, such trips are a convenient alternative to independent travel, although several good birding sites on the way will be missed.

Buses from Banjul, bound for Soma, pass the end of the laterite, but there is no public transport to the camp. The same problem will arise for anyone intrepid enough to attempt the journey by bush taxi from Serekunda, with the added disadvantage of a hot, cramped and thoroughly uncomfortable trip.

The ideal solution, though expensive, is a privately hired vehicle. Tourist taxis outside the main hotels offer day trips to Tendaba, but these give barely time for a quick look round and it is far better to strike a private deal with a taxi driver or with one of the Suzuki operators. A three-day trip with two overnights at the camp is ideal and gives time for plenty of stops on the way, but the driver's accommodation may have to be paid for.

Accommodation

The only place to stay is Tendaba Bush Camp, which is alongside the village. The accommodation is hardly luxurious and most of the bedrooms are round huts with no electric fans and with showers and toilets in a separate block. It is worth paying the extra D15 per person per night for a 'VIP' room with its own toilet, shower and ceiling fan. Many of the mosquito nets and screens have large holes in them, and all meals are served out of doors, so a good supply of insect repellent is essential. Be sure to bring enough water to last, since the only drinking water available is mineral water from the bar, and this is expensive.

Food is good and not too expensive, with a Scandinavian-style breakfast. There is a tourist buffet and a good á la carte menu.

The camp runs its own excursions, by pirogue to the north bank and by Landrover to Kiang West National Park. Some 1993 prices were (per person): overnight accommodation D135-150; breakfast D45; tourist buffet D90; á la carte D42-62.

Key

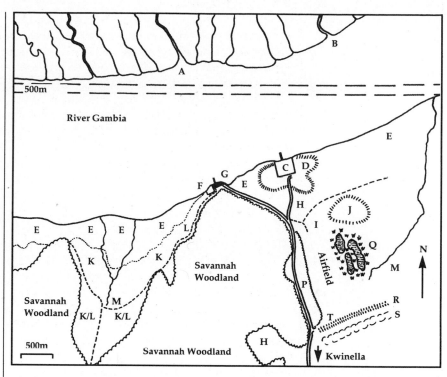

Duntu Malang Bolon	A
Tanku Bolon	B
Peanut Factory Compound	C
Konkoba Hill	D
Mangroves	E
Tendaba Bush Camp	F
Tendaba Village	G
Cultivations	H
Grassy Area	I
Low Wooded Hill	J
Rice Fields	K
Marshy Areas	L
Small Creek	M
Wooded Area	P
Swampy Area with some standing water	Q
High Earth Bank	R
Row of tall trees	S
Access to Airfield	T
Footpath (sometimes narrow)	- - -
Laterite road or track	===

It is not possible to book in advance since there is no telephone and no postal address but there are usually plenty of rooms available, although the 'VIP' accommodation is in demand and may be full in high season.

Strategy

Until about Christmas the tall elephant grass makes it difficult to follow some paths and restricts visibility to a few inches in places, but there is plenty of water around to attract birds. In October/November the bishops and whydahs are in breeding plumage. By New Year, everywhere is looking more open, but the rice fields have dried out and Yellow-crowned Bishops will be difficult to identify.

With private transport there are many good places to stop along the road and it is impossible to mention them all. The examples chosen can all be visited on a fairly leisurely drive (leaving the coast at about 7am and arriving in Tendaba in time for lunch) and include savannah woodland, rice fields, creeks, and the classic site of Brumen Bridge.

Just over 12km from the turn-off for Pirang, and shortly before the village of Kafuta, the road crosses a small tidal creek (Map Reference Sh11:402599), which eventually joins the Jaleh Cassa Bolon. On the right hand side of the road the end of the creek is surrounded by an easily accessible area of woodland in which African Pied Hornbill, Black Scimitarbill, White-crested Helmet-Shrike and African Yellow White-eye may be found.

About 10km further on, just before the village of Sutu Sinjang (Sh11:492582), there is another wet area, bordered by rice fields, where a variety of herons and egrets can usually be seen. From here to just beyond Bessi (Sh11:524590), where the Pima Bolon is crossed, there are several good places - creeks, rice fields and other cultivations - and a good number of raptors can be seen as the temperature rises and the thermals become established.

Brumen Bridge (Sh13:096650) is about 60km from Bessi and the area around is one of the classic birding sites in The Gambia, with Pel's Fishing Owl being one of the more exciting possibilities. Unfortunately, major work on the bridge has been causing considerable disturbance, and the current rate of progress suggests that it will continue to do so for some time. Even so, the area of mud and dead trees which borders the road beyond the bridge still attracts large numbers of waders, herons, egrets and other birds.

A final brief stop should be made at Kwinella (Sh13:135810), which lies at the junction of the main highway with the laterite to Tendaba. The village has nothing special to offer apart from the large roosts of Pink-backed Pelicans which occupy surrounding trees. This remarkable sight is easily visible from the Tendaba road.

Birds **The Camp area and the Rice Fields** – At the south-west corner of the camp compound a small gate leads onto a narrow track. The gate is open during much of the day but is padlocked in the late afternoon, necessitating a detour round the edge of the village. A right turn along this track leads towards the rice fields, with savannah woodland covering a slope to the left, and mangroves bordering the river to the right. Early in the season elephant grass makes passage rather difficult

Blue flycatcher

in places and restricts visibility, but after December the grass has gone and the area is much more open. At a fork in the path, bear right, towards the mangroves, to reach the edge of the mud, which is dry and firm even in October. There are good views across the rice fields to the high mangrove forest beyond, with a good chance of Woolly-necked Stork, Sacred Ibis and Blue-breasted Kingfisher as well as the commoner herons and egrets. The edge of the savannah is fringed with low scrub, and usually has several species of sunbirds, warblers and estrildine weavers, including Black-rumped Waxbill.

A footpath leads between the rice fields and mangroves on the right and the edge of the savannah on the left. About a mile from the camp it reaches a large open area containing rice fields, and crossed by a small creek where Spur-winged Goose may accompany the usual herons and egrets. Here one fork in the path bears right towards the river, while the other heads left towards the village of Batelling, some 3km away.

This second path leads along a low bank and passes a swampy area before entering a good stretch of savannah woodland, where Brown-necked Parrots may sometimes be seen, as well as Bruce's Green Pigeon, Brown-backed Woodpecker, African Yellow White-eye and Chestnut-crowned Sparrow-Weaver.

The path emerges into cultivations before reaching Batelling. It is best to turn round here and return to Tendaba by the same route.

The Airfield and its Surrounds – Leaving Tendaba village towards Kwinella, the road passes between trees and scrub on the hillside to the right, and rice fields which stretch towards the river on the left. On the river bank beyond the cultivations is Konkoba Hill, on which a peanut factory is situated and where Mosque and Red-rumped Swallows are commonly seen. About 1km down the road, a broad laterite track leads off left to the factory compound, and immediately beyond this, in the angle formed by the track and the Kwinella road, a narrow path leads into the scrub. During the first part of the season the path runs through tall elephant grass and it is easy to become disorientated, so a compass may be useful on this particular walk. The path soon forks, with the right branch leading south towards the airstrip and the other heading east past some cultivations towards a region of mixed scrub and grassland which lies roughly between Konkoba Hill to the left and another low, wooded rise to the right. The main swampy area to the east of the airstrip is visible from here, but is best approached from further down the main road. The tall grass hides many species in the early dry-season months, but there are places where visibility is much better. Abyssinian Ground Hornbills sometimes occur here but are much more likely to be seen when the grass has gone. Over to the left, in front of Konkoba Hill, lies an area of woodland, and just in front of this is a swampy area where herons and egrets congregate and where Blue-breasted Kingfishers may sometimes be seen. A narrow side-path leads towards it, but is not always easy to distinguish.

To reach the airfield itself, it is necessary to go about a mile further down the road towards Kwinella. Several narrow tracks lead off to the left, but should be ignored. The approach to the airstrip is at its southern end and is wide enough to drive a vehicle down. It is actually possible to drive along the whole length of the airfield from this point.

The airstrip is still used by an occasional light aircraft during the dry season, but there is nothing to mark the actual runway which is merely a long, wide area of dry mud and low grass, separated by trees and scrub from the laterite road on the west side and bordered on the east by an expanse of tall grass with swampy areas and some standing water during the first part of the season.

A steep earth bank forms the southern boundary of the airfield, and beyond this lies a row of tall trees which, when they are fruiting, are a good place to look for Bruce's Green Pigeon. Good views across the swampy expanse to the east can be had from various points along the airstrip and may reveal Yellow-billed Stork, Knob-billed Duck, Spur-winged Goose, and the occasional Black Crowned-Crane, while Plain-backed Pipit and Quailfinch also occur in the area and male Yellow-crowned Bishops stand out brilliantly early in the season. The trees and scrub between the runway and the road are usually alive with birds, with Striped Kingfisher and Bearded Barbet seen regularly as well the occasional Senegal Batis and a variety of warblers and estrildine weavers.

The Creeks on the North Bank – The north bank of the Gambia River, opposite the camp, is studded with creeks, some narrow and impassable, others wide enough to permit the passage of a fair-sized boat. Pirogue (native-style motorized canoe) trips can be arranged from Tendaba Camp but may be expensive for a small party. The charge per person was D90 in 1993, but if there are less than four occupants then the charge is for the pirogue itself (D360). A better deal may be found in the village, but always agree a price before starting out. Always check that the pirogue is large enough to take the whole party and its equipment in safety as there are stories of birdwatchers, festooned with cameras, telescopes and binoculars, starting to sink gracefully in mid-stream because the boat had been overloaded by its avaricious owner. If you are planning to take one of the camp's own pirogue excursions, and you are not going to be the only group on board, make sure that any others are also birdwatchers as the boats can be hired for other purposes.

The main creek visited by the camp excursions is Tanku Bolon, where it is possible to disembark and walk along a path to a small village where African Swallow-tailed Kites may sometimes be seen. Another popular inlet for birdwatchers is Kisi Bolon, which is not marked on any official map and which has a narrow and inconspicuous entrance. Make sure in advance that the boatman understands exactly where to go and why, or the trip may become an expensive disappointment.

The birds present may depend on the state of the tide - at low tide, the stilt roots of the Red Mangroves provide perches for the Blue Flycatchers, while the best chance of a African Finfoot is just after high tide, when they seem to be flushed out into more open water.

Other possibilities on a trip to the north bank include Great White Pelican, Goliath Heron, Collared Pratincole, Blue-breasted Kingfisher, Northern Carmine Bee-eater and Mouse-brown Sunbird.

Other wildlife

Apart from the three common species of monkey, Western Baboon occur in the area. Spotted Hyaena, Aardvark and Cape Clawless Otter also occur but are unlikely to be seen. Warthog are common (the meat, worth trying, features prominently on the Tendaba Camp menu as bush pig). The Lesser Galago, a bushbaby, is found in the surrounding savannah, and is best looked for at night when a torch may reveal one clinging to a branch. They have very bright eyeshine and can be spotted at quite a distance.

Barra and Essau

The town of Barra is rather uninspiring, but the area around Fort Bullen, the ruined 19th century British fort, and the small lagoon with its associated creek, offer good birdwatching opportunities, with the chance of seeing some species, such as Cut-throat, which occur rarely, if at all, on the south bank of the Lower River.

The village of Essau can claim to be the closest point to the resort areas where Northern Anteater-Chats are regularly seen. The surrounding parkland savannah can also be easily explored on foot and harbours several other species which are unlikely to be seen south of the river.

Location

Map References: Barra (Fort Bullen) - Sh10:322914; Essau (Road Junction) - Sh10:337909. The town of Barra lies immediately opposite Banjul on the north bank of the Gambia River estuary. The village of Essau is situated just over 1km east of Barra, at the point where the laterite road to Berending and Farafenni branches off from the main asphalt road which swings northwards towards the Senegal border, 16km or so away.

The estuary is crossed by car ferries operating between Banjul and Barra. These ferries carry foot passengers and the fare increased to D5

Key

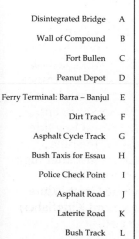

Disintegrated Bridge	A
Wall of Compound	B
Fort Bullen	C
Peanut Depot	D
Ferry Terminal: Barra – Banjul	E
Dirt Track	F
Asphalt Cycle Track	G
Bush Taxis for Essau	H
Police Check Point	I
Asphalt Road	J
Laterite Road	K
Bush Track	L

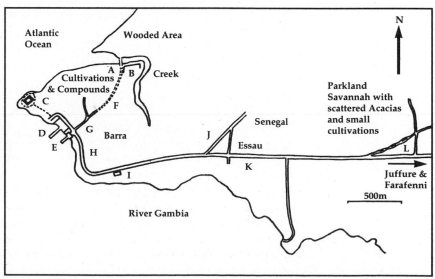

at the beginning of the 1992/93 season. This is for a single journey as return tickets are not available. The Banjul terminal is towards the end of Wellington Street, about three blocks down from the African Heritage Restaurant (see map, p10). The first sailing from Banjul to Barra is at 08.00, with two-hourly sailings from then on. Return trips from Barra to Banjul are also two-hourly from 09.00 to 19.00. However, ferries are rarely on time and the timetable is liable to change without notice or be conveniently ignored, so it is unwise to rely on the last ferry back. Tickets are purchased in advance from the ticket office and not (as some guide books suggest) on board the boat. Turn off Wellington Street towards the terminal, there some large double gates give access to the quayside for vehicles. On the left of these is the passenger waiting room; a narrow passageway runs down the back of it and the ticket office is located along this. Arrive at least 15 minutes in advance of the published departure time, even though the ferry will probably be late, since crossings are usually overcrowded and fairly chaotic and it is a good idea to be near the front of the queue.

The crossing takes half an hour or so and it will be worthwhile trying to occupy a good vantage point, since dolphins sometimes put in an appearance and there are usually at least a few terns around, with good numbers of White-winged Black Terns on passage in April. On reaching Barra, check the return ferry sailings on the board outside the ticket office and aim to catch the penultimate one, which should be at 17.00.

All the birdwatching sites around Barra are within a short walk of the ferry terminal. From Barra it is a short trip in a bush taxi to the main junction in Essau. Make it clear to the driver that this is where you wish to disembark otherwise he will head rapidly towards Senegal.

At the ferry terminal at Barra drivers will offer to take you here, there and everywhere, but at tourist rates. It is better to turn right immediately outside the terminal building and walk about a hundred metres up the road to where the 'tankatanks' are parked.

The road passes through dilapidated suburbs for just under 1km and then reaches a police check point where the driver should pull in and pay his fee. Essau lies about the same distance beyond this, at the point where the only wide laterite road branches off to the right into the main part of the village, the junction being marked by a tree of impressive size.

To catch a bush taxi back to Barra, it will be necessary to stand by the side of the asphalt road a little way down from this and look out for one coming down from the border.

Strategy   The unpredictability of transport makes it best to get the 8.00am ferry crossing to Barra and go at once to Essau, spending the morning in the savannah. This leaves plenty of time to explore the Barra area during the afternoon and catch the penultimate ferry back to Banjul. December is a good time to visit Essau, for then the cultivations have mainly been cleared and birds such as Chestnut-backed Sparrow-Lark will be that much easier to spot, while Acacias are in flower and attract the usual good variety of species.

Northern Anteater
Chat

Birds   From the road junction in Essau it is a walk of about 1500m through
the village and out into the savannah. When birdwatching in the village
itself, you may find curious locals, especially children, can be a
nuisance but they are good natured and no real problem. Northern
Anteater-Chats can turn up in the village itself, as well as further on
along the roadside, and there is also the chance of a Chestnut-bellied
Starling.

          After about 800m, the buildings start to thin out and the road climbs
perceptibly as it nears the open countryside. The last building is on the
left hand side, and then, on the brow of the hill, a well-defined sandy
bush track heads off to the left, through parkland savannah with
scattered cultivations. The track, which divides and rejoins a number of
times, eventually reaches the village of Jiffet from where another track
rejoins the main road. It is only necessary to walk a short distance to an
area of scattered Acacias and other trees which attract a wide variety of
species, especially when flowering or fruiting. African Green Pigeon,
Eurasian Hoopoe, Scarlet-chested Sunbird, Pin-tailed Whydah, Cut-
throat, Chestnut-crowned Sparrow-Weaver and Speckle-fronted
Weaver are among the possibilities. Striped Kingfisher, Brown-backed

Woodpecker and White-rumped Seedeater are also found here and Chestnut-backed Sparrow-Larks occur regularly in the cultivations. There are plenty of raptor perches with a good chance of a Dark Chanting Goshawk and an overflying African Fish Eagle is possible.

To explore Barra it is best to turn left immediately outside the ferry terminal and walk along the back of the peanut factory compound where there may be a wagtail or two, and occasionally a African Silverbill or a Village Indigobird. At the far end of the compound, the track turns sharp left, back towards the estuary, and the entrance to the fort is a few yards down here on the right. You only have to pay to visit the ruin itself, and not the surrounds which are mainly fairly open ground with scattered trees and shrubs. African Green Pigeons occur here when the trees are fruiting, and there is also the chance of Mottled Spinetail, Vieillot's Barbet and possibly a Cut-throat.

From the fort it is possible to walk along the shore, where a sand-bank cuts off a narrow lagoon. Here there are usually several species of waders, as well as a variety of gulls and terns, while the sparse grass bordering the shore is home to Crested Larks.

Follow the creek around to the right to the remains of a wooden bridge which ends in mid-stream. Walk beyond the bridge, along the edge of the mangroves, as views can be had of the far side and a number of species frequent the area, including warblers, Yellow-crowned Gonolek and White-billed Buffalo Weaver. From the bridge, a well-defined track leads back through fairly open ground with patches of low scrub before entering the town again and finally ending immediately opposite the ferry terminal.

Basse Santa Su

The highlight of any trip to Basse, as it is commonly known, must be its Egyptian Plovers, which are regularly present between November and the end of January. However, as well as these beautiful birds, a number of other Middle and Upper River specialities, such as Red-throated and Northern Carmine Bee-eaters, Swamp Flycatcher, Cut-throat and Long-tailed Paradise Whydah may also be found, especially in the areas bordering the Prufu Swamp to the east of the town. African Mourning Doves, rare on the coast, are common in Middle and Upper River regions where they largely replace the very similar Red-eyed Doves.

Location

Map Reference - Sh20:850715. The town of Basse, which is the capital of the Upper River Division, lies at the eastern limit of the tarmac road south of the river, some 385km from Banjul. Buses leave Banjul (Cotton Street) in the early morning to drive direct to Basse, but the jouney takes a good nine hours and is only recommended for the really desperate. As published timetables seem non-existent, check with the GPTC main depot at Kanifing for times. West African Tours, Gamtours and Black-and-White Safaris often run trips to Basse with various

Key

Ferry Jetty A

Marshy Area with Pools B

Chamoi Bridge C

Roads (Asphalt or Laterite) ═══

Tracks ≡≡≡

Selected Footpaths - - -

stopovers on the way, but the tour itinerary is not specifically designed for birdwatchers. It is also possible travel by bush taxi from Serekunda to Soma and then on to Basse, but your condition on arrival is best left to the imagination.

Private hire is the ideal solution, giving the opportunity to visit places such as Tendaba and Jakhaly on the way as well as the chance of a more thorough exploration of the Basse area. In addition to these possibilities, Lamin Touray, the Gambian partner of "Dreambird", is happy to take people along on trips if there are spare places, usually visiting Tendaba, Georgetown and Basse. His telephone number in The Gambia is 82156 and if unable to fit in any 'extras' himself he may be able to arrange transport with someone reliable.

Accommodation

Although there are three hotels in Basse - the Plaza, the Teranga and the Apollo 2 - the best plan is to stay at Janjangbure Camp (sometimes known as Peter's Place) on the north bank of the river opposite Georgetown. This is a fairly new safari camp, along the lines of Tendaba but rather more comfortable and slighty more expensive. Unlike Tendaba it is possible to book accommodation in advance, through Gambia River Excursions, PO Box 664, Banjul (Tel. 95526). From Georgetown to Basse is about 75km, so an early morning start will give plenty of time for a good exploration of the area before returning.

Strategy

Egyptian Plovers can be seen at Basse throughout the first part of the dry season, with maximum numbers usually present in November, decreasing slowly from then until the end of January by which time they have normally gone. The Prufu Swamp area (see above) and Jakhaly (see p.86) are also best visited early in the season while there is still plenty of water around. By early spring many areas have dried out and birdwatching can be poor.

If possible, it is a good idea to combine a trip to Basse with one or two nights at Tendaba, since not only does this provide a welcome break from the apalling road (which does not improve until after Soma) but it also gives the chance of some birdwatching stops on the way. In particular, a visit to Jakhaly can be worthwhile if time allows, since although the famous freshwater swamp has now been drained and replaced by rice fields, and the large flocks of wildfowl and waders have departed, it is still possible to find African Pygmy Goose, African Crake, Allen's Gallinule and Greater Painted-snipe.

Jakhaly lies just off the main road about 29km west of Georgetown (Map Reference - Sh8:029975) and the rice fields, which cover an extensive area to the north-east, are best approached through the village. About 1500m after the turn-off from the main road, tracks branch to the left and right around the edges of the rice-fields which are crossed by drainage ditches (see map). In both directions the tracks run into woodland which borders the river and is rich in birdlife. The tracks actually link up, providing a circular walk of some 13km. The area around the main drainage canal, and especially the region between the track and the river, seems a particularly rewarding place for birds, as does the swampy area to the west of this.

The roadsides in Middle and Upper River should be watched for a number of species which are rare or absent nearer the coast. Rüppell's Griffon Vulture is not uncommon and often to be found at carcasses, along with the commoner White-backed Vulture, while the scarce White-headed Vulture may also sometimes be seen. Marabou Storks are locally common, especially along the Jappeni-Georgetown-Bansang stretch of the road, where there are numerous colonies in Baobabs, mainly occupied from October to March but with individuals often still present later in the season.

Key

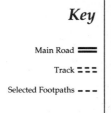

Main Road ▬▬▬

Track ▬ ▬ ▬

Selected Footpaths ▬ ▬ ▬

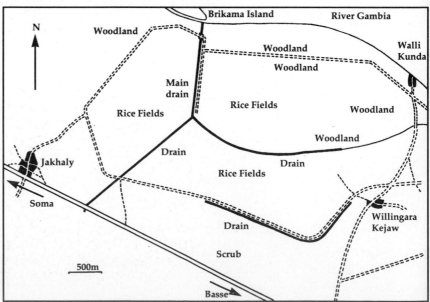

Birds

The river at Basse is crossed by a passenger ferry, and it is from the south bank ferry jetty that the best views of Egyptian Plover are usually obtained. About 200m inland of this jetty, a bush track heads almost due east towards the village of Damfa Kunda which lies just over 4km away on the other side of the Prufu Swamp. Three bridges carry the track across a series of creeks and there are plenty of pools and swampy areas around during the first part of the dry season. Good views over the creek and waterside vegetation can be had from the first bridge and an early morning walk should turn up a good number of species. The surrounding scrub harbours Stone Partridge as well as Red-throated Bee-eater and Pygmy Sunbird, with a good chance of Cut-throat, and Northern Carmine Bee-eater is seen regularly in the vicinity of the river. Just before the first bridge a path leads along the creek to a series of pools on the edge of a marshy area. Black Crake, African Jacana, Greater Painted-snipe, Blue-breasted Kingfisher, Swamp Flycatcher and Long-tailed Paradise Whydah may all be found in this area, with Verreaux's Eagle Owl also having been recorded. The stream has its origins in the Prufu Swamp and cannot be followed for much further, although the area is good for raptors. About 400m from the bridge, another track joins the path and provides a convenient return to Basse.

Leaving Basse south on the main road towards Sabi and the Senegal border, turn left at Mansajang Kunda (just over 1km) on the laterite road to Fatoto. After 4.5km, not far from the village of Chamoi, a bridge carries the road over the Prufu Bolon (Map Reference Sh20:900/16) and here a path runs along the edge of the creek on both sides of the road. In the early part of the season it is impassable in places, but even a short walk may turn up Hadada Ibis, Greater Painted-snipe, or Long-tailed

Long-tailed Paradise
Whydah

Paradise Whydah, while Wahlberg's Eagle is also recorded from the area. The path on the south side of the road is especially rewarding, with Grey-headed Kingfisher, Red-throated Bee-eater, Swamp Flycatcher and Bronze-tailed Glossy Starling among the many possibilities.

Other wildlife   The most exciting possibility in the Middle and Upper River is the chance of seeing Hippopotamus, which are still not uncommon from Elephant Island eastwards, although they are more likely to be sighted during a river trip.

Quite apart from the pirogue trips to the north bank at Tendaba, several other opportunities exist to birdwatch from a boat on the river.

Most of the major tour companies offer a variety of excursions geared to the requirements of the average holidaymaker. The Creeks Tour and Champagne and Caviar excursions are not deliberately aimed at birdwatchers but both spend much of the time slowly cruising the bolons off the Gambia River, including Oyster Creek, and offer the chance of seeing some good birds.

Lamin Lodge

Lamin Lodge is situated about a mile outside Lamin village, on the bank of the Lamin Bolon. It is a rustic, 'African-style' bar and restaurant, built on stilts among the mangroves and overlooking the creek. The restaurant caters especially for birdwatchers and provides early-morning 'birders' coffee' from 07.00 and breakfast from 08.00. A substantial buffet meal and an á la carte menu are both available until 20.00 when the restaurant closes. Dugout canoes, with or without engines, can be hired by the hour to explore the creeks, and for anyone unable to make the Tendaba trip, this presents a good opportunity to see Goliath Heron and other species. Larger motorised pirogues with all facilities are available for larger parties, with what is described as 'adventurous overnight accommodation on board' for fifteen persons.

The track down to the Lodge is on the left hand side just as you enter Lamin village from the Serekunda direction and is well-signposted. It takes about twenty minutes to walk down to the creek through a good birdwatching area, so a visit could easily be combined with a trip to Abuko, which is only a short distance away. Prices (1993) were: Breakfast D35; Buffet D60; Large pirogue (for parties) D1,500 per day; Dugout canoe D50 (per hour); Outboard Engine D150 (per hour). Bookings may be made in advance, and further information is available from Samba River Venture Ltd, PO Box 664, Banjul, The Gambia (Telephone 95526)

Finally, it is possible to take a three day river cruise from Basse to Banjul on the M.V. El Ouata III, with plenty of interesting birdwatching en route. Cost of the trip is about D2000 per person and details can be obtained in The Gambia by phoning Jane Martin on 91883 or by writing to Clive Barlow, PO Box 296, Banjul, The Gambia.

SELECTIVE BIRD LIST

White-backed Night-Heron. Resident but rare and very local. Has bred in Abuko (p24) and was seen regularly in the Gumbar to the right of the Education Centre balcony in 1992.

Woolly-necked Stork. Probably resident in Lower and Middle River regions but rare near the coast. Occurs regularly on the muddy north bank of the Bintang Bolon at Brumen Bridge, and around the rice fields to the west of the camp at Tendaba (p76).

Sacred Ibis. A rare resident, found mainly in the Middle and Upper River regions. Occurs at Tendaba (p76) and sometimes near the coast during the dry season. Recently seen at Camaloo Corner (p52) and the Bund Road (p55).

African Spoonbill. A rare resident and dry season visitor. Found in Lower and Middle River and regularly recorded from the Bund Road (p55). Also occasionally at Camaloo Corner (p52).

African Pygmy Goose. Rare. Mainly a Middle River resident and found on freshwater marshes. Still seen at Jakhaly (p86) even though the large swamp has now been drained and converted to rice fields.

African Swallow-tailed Kite. A rare dry season visitor to Middle and Upper River and very rarely near the coast (one appeared at Camaloo Corner in January 1993). Sometimes seen in the savannah on the north bank of the river opposite Tendaba Camp (p76).

Bateleur. A rare resident, formerly much more common. Recorded regularly from Kiang West National Park (near Tendaba) and also from the Finto Manereg area (p69) where a pair was seen frequently from the Faraba Banta-Jiboroh Kuta bush track from October to mid-December 1993.

Gabar Goshawk. Probably a rare resident, although breeding has not been proved. Occasionally seen in the Finto Manereg area (p69) and along the Sukuta-Tanji road.

Western Little Sparrowhawk. A rare resident in open woodland in the Lower and Middle River regions. Regularly recorded at Abuko (p24).

African Goshawk. A rare forest resident recorded regularly from Abuko (p24), where it probably breeds, and also from Tanji, Brufut and Pirang.

Wahlberg's Eagle. A rare resident, found in all regions where there are tall trees. Recorded from the Bessi area on the Banjul-Tendaba road, and from Basse (near Chamoi) (p84).

Tawny Eagle. A rare resident, mainly in Lower and Middle River, and most commonly found in 'parkland savannah' and other areas with scattered trees. Regularly seen in the Pirang-Finto Manereg-Seleti area.

African Hawk Eagle. Resident in open forest but rare in all regions. Occurs in the tall woodland around Seleti (p73).

Martial Eagle. Resident and recorded from all regions but rare everywhere. Regularly seen in the Finto Manereg area (p69) - a male perched in a tree by the side of the track in November 1992.

Red-necked Falcon. A rare resident, usually found in areas where there are Rhun Palms. A pair was seen regularly in the Kotu-Fajara area during 1992/93.

White-spotted Flufftail. A rare resident in gallery forest and wet scrub but extremely shy and difficult to see. Regularly recorded from Abuko (p24) and observed nest building at Pirang in 1991.

African Crake. A rare resident, found in all regions where there are freshwater swamps but difficult to see because of its skulking habits. Regularly recorded from the Bund Road (p55), Camaloo Corner (p52), Abuko (p24), Jakhaly (p86) and the Prufu Swamp at Basse (p85).

Black Crowned-Crane. An uncommon and local resident, mainly in Middle River but may sometimes be seen at Brumen Bridge and in the marshy area bordering the airfield at Tendaba (p76). The most reliable place to see them nearer to the coast is at Pirang (p64), where they regularly roost in the mangroves.

African Finfoot. Resident and recorded all along the river from Kotu to Basse, but an extremely elusive bird and only very occasionally seen. Regularly recorded at Abuko, but most frequent sightings are in the creeks on the north bank of the river at Tendaba (p76), especially just after high water, when they seem to be flushed out of the mangroves where they usually skulk.

Greater Painted-snipe. Rare and mainly a dry season visitor but recorded throughout the year wherever there are swampy areas. Seen regularly in late season at Camaloo Corner (p52) and also recorded from the Bund Road (p55), Jakhaly (p86) and Basse (Prufu Swamp) (p85). During November/December 1992 was frequently seen around the periodically wet area behind the Palma Rima and in the swampy ground behind the Badala Park Hotel at Kotu (p39).

Egyptian Plover. Mainly a dry season visitor to the Upper River. Regularly seen at Basse (p84) during the first half of the dry season, when they can usually be viewed from the jetty which serves the ferry on the south bank.

Temminck's Courser. An uncommon resident in open country, most often seen in cultivated areas after they have been cleared in December. Regularly found around the airport at Yundum where the fields bordering the road behind the police check point are worth scanning. Also in cultivations behind the Agricultural College Complex (p60), and in the water melon fields behind the womens' fields at Lamin (p57).

Four-banded Sandgrouse. Rare in the Lower river region although locally common in dry open bush further inland. Occurs in scrub at Pirang (p64) and can be seen regularly at the water-holes at Seleti (p73) when they come in in good numbers to drink at dusk.

Brown-necked Parrot. A rare and local resident, breeding in the areas of tall Red Mangrove which occur on the south bank of the river from Pirang east to just beyond Elephant Island. Flocks are sometimes seen flying over the mangroves at Pirang (p64) and on the edge of the savannah at Tendaba (p76).

Violet Turaco. Resident and found throughout the country but uncommon and restricted to the small areas of remaining forest. Seen regularly at Bijilo (p29), Tanji (p33), Brufut (p37) and Finto Manereg (p69), in the tall woodland around Seleti (p73), and especially at Abuko (p24).

Guinea Turaco. Resident but rare and very local in the residual high forest of the Lower River. Recorded from Pirang, Brufut, Tanji and Seleti but most commonly seen in Abuko (p24). Can be difficult to see as it often sits motionless and well-camouflaged, high in the tall trees, until it takes wing and reveals a flash of brilliant crimson.

Mottled Spinetail. Rare and local resident, especially in Middle River but also along the coast. Regularly recorded from Brufut (p37) and also from Seleti (p73) and Barra (p81).

Blue-breasted Kingfisher. An uncommon resident, found mainly along the river in the dry season, particularly in Middle and Upper regions, but one was fishing at the large water hole at Seleti (p73) in November 1992. Seen regularly in the Basse area (p84) and at Tendaba (p76), on the edge of the ricefields and the marshy areas around the airstrip. In late 1992 there was a pair at the Katchikally Sacred Crocodile Pool in Bakau.

Giant Kingfisher. Resident on the coast and inland, both along the river and near creeks and deep pools, but nowhere common. A pair is regularly in evidence in the Kotu Creek area (p39), often on the telegraph wires. Also seen regularly at the Crocodile Pool at Abuko (p24) and the lagoon at Tanji (p33).

Red-throated Bee-eater. Mainly a non-breeding dry season visitor to Middle and Upper River, although there have been sporadic records of breeding. Not found in the Lower River region. Frequently seen in the Basse (p84) area where breeding has occurred on a regular basis.

Northern Carmine Bee-eater. Scarce in the Lower River region but a common dry season visitor further east, with regular sightings from the north bank of the river at Tendaba (p76) and in the area around Basse (p84).

Black Scimitarbill. A rare and local resident wherever there is suitable woodland. Recorded from Bijilo (p29), Tanji (p33), and the wooded area which borders the creek near Kafuta.

Abyssinian Ground Hornbill. Resident but uncommon throughout the country. Usually found in cleared cultivations or in open bush with scattered trees. Regularly seen in the Faraba Banta-Jiboroh Kuta-Madina Ba area (p68), especially once the cereal crops have been harvested in December.

Yellow-rumped Tinkerbird. A rare resident in woodland in Lower River. Reported as regularly seen at Fajara and Brufut, it is also one of the scarcer species to be found at Abuko (p24).

Spotted Honeyguide. A rare forest resident from the Lower River and further inland. Regularly seen at Abuko (p24).

Buff-spotted Woodpecker. A rare resident in forest in the Lower River. Regularly seen at Abuko (p24), where the area of gallery forest between the Education Centre and marker post 24 is a good place to look. Also reported from Brufut and Pirang.

Cardinal Woodpecker. A rare and local resident in clearings, on woodland edges, and in areas with groups of trees. Occurs at Bijilo (p29), Yundum (p59), Tanji (p33), and in the area of the Casino Cycle Track (p43) and the Old Cape Road (p52). Also recorded from Tendaba (p76).

Brown-backed Woodpecker. Resident in more open woodland throughout the country but rare everywhere. Recent records include Brufut (p37), Seleti (p73), Kabafita (p62), Essau (p81) and Tendaba (p76).

Chestnut-backed Sparrow-Lark. A local resident, not uncommon in open bush on the north bank in Middle and Upper River, but rare elsewhere. Occurs regularly at Yundum (p59) and Pirang (p64), but most likely to be seen in the Basse area (p84) and on the edge of the bush at Essau (p81), especially in December around cleared cultivations.

Yellow-throated Long-claw. An uncommon resident on the edges of freshwater swamps and marshes in all regions. Recorded from Camaloo Corner (p52), Abuko (p24) and Jakhaly (p86). Regularly seen in short grass bordering the mudflats around Cape Creek by the side of the Old Cape Road (p52).

White-breasted Cuckoo-shrike. A rare and local resident of remnant forest, mainly in Lower River. Seen regularly at Kabafita (p62) and also found at Yundum (p59), Tendaba (p76), and in the tall forest near the Seleti water-holes (p73).

Grey-headed Bristlebill. Another rare and local forest resident, regularly recorded at Abuko (p24) and also occurring at Brufut and Pirang.

Northern Anteater-Chat. A regular dry season visitor and possibly a local resident on the north bank of the river. Rarely seen on the south bank. Seen regularly in and around the village of Essau (p81), which is the most reliable place to find one near to the coast.

White-fronted Black Chat. Found in all regions as a rare and local resident in clearings and areas of open woodland. Regularly reported from the area between the Yundum Agricultural College and the village of Old Yundum (p59).

Green Crombec. A recent addition to the Gambia list. Recorded from several sites including Tanji, Brufut and Pirang, and appears to be a local breeding resident, favouring trees covered with a dense growth of creepers and moving around in the foliage of dense thickets and the canopy down to near ground level.

Green Hylia. A rare and local forest resident with almost all sightings being in the coastal area on the south bank of the River Gambia. Regularly seen in Abuko (p24), with other recent sightings including Brufut and Pirang.

Yellow-breasted Apalis. Found in the Lower and Middle River as a rare and local resident of the remaining forest. Regularly seen at Abuko (p24), and recently recorded from several new sites including Brufut, Pirang, and Tanji.

Pale Flycatcher. A rare resident of open woodland mainly in the Lower River region. Reported regularly from Yundum (p59), where it breeds.

Lead-coloured Flycatcher. A rare and local resident of remaining high forest throughout the country. Reported sightings from Tanji, Pirang, Kabafita, Tendaba, Jakhaly and Basse, and more recently in the woodland around the Seleti water-holes (p73).

Senegal Batis. Local and uncommon resident in Lower River woodland. Regularly seen in Bijilo Forest Park (p29) and at Kabafita (p62).

Blue Flycatcher. Probably resident, and occurring very locally along the forest edge from Tendaba to the east of the country. Seen regularly on the north bank of the river opposite Tendaba (p76), especially in creeks such as Tanku and Kisi Bolons where individuals often perch on the exposed stilt roots of the Red Mangroves at low water.

White-winged Black Tit. A rare and local resident of open woodland, mainly in the lower River region. Small parties recently seen in the woodland at Seleti (p73) and on the edge of the Finto Manereg Forest Park near Douassu (p69).

Yellow Penduline-Tit. A rare resident of open forest recorded from Brufut and also from Yundum (p59) with several sightings behind the Agricultural College.

Mouse-brown Sunbird. A rare resident in the mangroves and adjacent trees from the coast to Tendaba (p76) where individuals are frequently seen on the north bank, particularly in the creeks.

Collared Sunbird. A rare forest resident of the Lower River region recorded at Fajara and Kabafita and especially at Abuko (p24), where individuals are regularly seen in the trees to the left of the Education Centre balcony.

Pygmy Sunbird. A rare resident, found in dry bush and woodland, especially inland. Has been seen at Essau (p81), Jakhaly (p86) and especially in the Basse area (p84).

Green-headed Sunbird. A rare Lower River woodland resident. Occurs in the Bijilo Forest Park (p29).

African Yellow White-eye. A rare and local resident in open woodland throughout the country. Recorded from Yundum, Kabafita, Tendaba and Jakhaly and recently from the woods bordering the creek near Kafuta. At the Finto Manereg Forest Park near Douassu (p69) several individuals have been seen feeding in trees on the edge of the bush track together with barbets, orioles and estrildine weavers (December 1992).

Sulphur-breasted Bushshrike. A rare resident of dense scrub and wooded areas. Recorded from Brufut (p37), Fajara (p46), Abuko (p24) and Yundum (p59) with recent sightings at Tanji (p33) and Kabafita (p62).

Speckle-fronted Weaver. Regular but uncommon dry season visitor to the north bank of the river where it occurs locally in dry, open savannah such as that at Essau (p81).

Chestnut-crowned Sparrow-Weaver. A rare and local resident in Lower and Middle River regions, found in open woodland and 'parkland savannah' where there are scattered groups of trees. Recorded from Yundum (p59), Essau (p81) and Tendaba (p76).

Western Bluebill. A rare and local resident of dense savannah woodland in the Lower River region, south of the river. Breeds in Abuko (p24), where it is regularly recorded. Seen recently at several new locations including Bijilo Forest and Pirang.

Black-faced Firefinch. A rare woodland resident in Lower and Middle River regions, recorded from Yundum, Tendaba and Jakhaly, and recently from the edge of the Finto Manereg Forest Park near Douassu (p69).

African Silverbill. Resident in both Lower and Middle River regions, mainly near the coast, but rare everywhere. Has been seen near the Peanut Depot at Barra (p84), along the Casino Cycle Track (p43), and regularly on the telegraph wires near the Paradise Beach Bar at Kotu.

Cut-throat. A rare dry season visitor to open bush and 'parkland savannah', mainly in Middle River. Also recorded from scrub on the edge of the Prufu Swamp at Basse (p84), from Pirang (p64) from the area around Fort Bullen at Barra (p81), and regularly from the bush around Essau (p81), where a small flock was feeding with other estrildine weavers in a group of Acacias in January 1993.

Long-tailed Paradise Whydah. Mainly confined to Middle and Upper River regions where it is a rare resident in open bush. May be found in scrub on the edge of the Prufu Swamp at Basse (p84) and also in the Chamoi area (p85).

The list includes all those species which have been reliably recorded in The Gambia, but excludes any unsubstantiated sightings. The taxonomic sequence and scientific nomenclature follow Dowsett & Forbes-Watson (1993). However, as a new field guide to the birds of West Africa which is currently being written is bound to become the main field guide used by birdwatchers visiting The Gambia the English names used are from the provisional list for that book, which has kindly been made available by Iain Robertson, one of the book's authors. Because the frequent use of alternative vernacular names often leads to confusion, a list of the more commonly encountered synonyms is appended.

KEY TO CHECKLIST

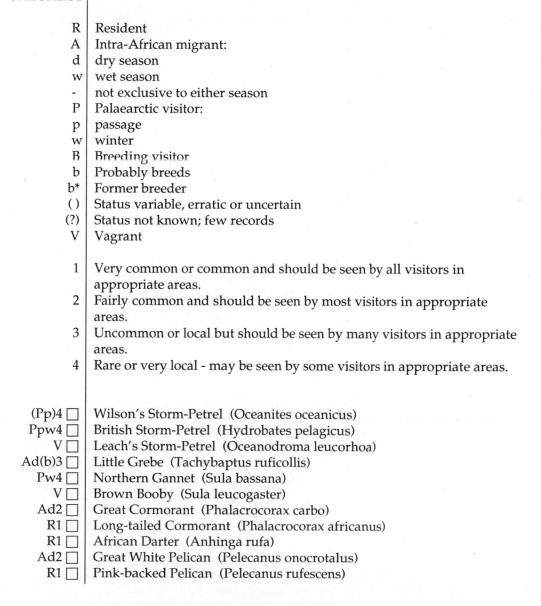

R	Resident
A	Intra-African migrant:
d	dry season
w	wet season
-	not exclusive to either season
P	Palaearctic visitor:
p	passage
w	winter
B	Breeding visitor
b	Probably breeds
b*	Former breeder
( )	Status variable, erratic or uncertain
(?)	Status not known; few records
V	Vagrant
1	Very common or common and should be seen by all visitors in appropriate areas.
2	Fairly common and should be seen by most visitors in appropriate areas.
3	Uncommon or local but should be seen by many visitors in appropriate areas.
4	Rare or very local - may be seen by some visitors in appropriate areas.

(Pp)4 ☐	Wilson's Storm-Petrel (Oceanites oceanicus)
Ppw4 ☐	British Storm-Petrel (Hydrobates pelagicus)
V ☐	Leach's Storm-Petrel (Oceanodroma leucorhoa)
Ad(b)3 ☐	Little Grebe (Tachybaptus ruficollis)
Pw4 ☐	Northern Gannet (Sula bassana)
V ☐	Brown Booby (Sula leucogaster)
Ad2 ☐	Great Cormorant (Phalacrocorax carbo)
R1 ☐	Long-tailed Cormorant (Phalacrocorax africanus)
R1 ☐	African Darter (Anhinga rufa)
Ad2 ☐	Great White Pelican (Pelecanus onocrotalus)
R1 ☐	Pink-backed Pelican (Pelecanus rufescens)

V ☐	Magnificent Frigatebird (Fregata magnificens)
V ☐	Great Bittern (Botaurus stellaris)
R/(Pw)4 ☐	Little Bittern (Ixobrychus minutus)
(R)4 ☐	Dwarf Bittern (Ixobrychus sturmii)
R/Pw1 ☐	Black-crowned Night-Heron (Nycticorax nycticorax)
R4 ☐	White-backed Night-Heron (Gorsachius leuconotus)
R/Pw2 ☐	Squacco Heron (Ardeola ralloides)
R1 ☐	Cattle Egret (Bubulcus ibis)
R2 ☐	Green-backed Heron (Butorides striatus)
R/Ad2 ☐	Black Egret (Egretta ardesiaca)
R/Pw2 ☐	Little Egret (Egretta garzetta)
R1 ☐	Western Reef-Egret (Egretta gularis)
R/Ad4 ☐	Intermediate Egret (Egretta intermedia)
R1 ☐	Great White Egret (Egretta alba)
R/Pw4 ☐	Purple Heron (Ardea purpurea)
Pw/A-1 ☐	Grey Heron (Ardea cinerea)
A-B2 ☐	Black-headed Heron (Ardea melanocephala)
R2 ☐	Goliath Heron (Ardea goliath)
R1 ☐	Hamerkop (Scopus umbretta)
A-B2 ☐	Yellow-billed Stork (Mycteria ibis)
V ☐	Black Stork (Ciconia nigra)
(A-)4 ☐	Abdim's Stork (Ciconia abdimii)
R4 ☐	Woolly-necked Stork (Ciconia episcopus)
V ☐	White Stork (Ciconia ciconia)
(R)4 ☐	Saddle-billed Stork (Ephippiorhynchus senegalensis)
A-B3 ☐	Marabou Stork (Leptoptilos crumeniferus)
R4 ☐	Sacred Ibis (Threskiornis aethiopicus)
Ppw4 ☐	Glossy Ibis (Plegadis falcinellus)
R3 ☐	Hadada Ibis (Bostrychia hagedash)
V ☐	Eurasian Spoonbill (Platalea leucorodia)
R/Ad4 ☐	African Spoonbill (Platalea alba)
(Pw/A-)3 ☐	Greater Flamingo (Phoenicopterus ruber)
V ☐	Lesser Flamingo (Phoenicopterus minor)
V ☐	Fulvous Whistling-Duck (Dendrocygna bicolor)
R1 ☐	White-faced Whistling-Duck (Dendrocygna viduata)
V ☐	Egyptian Goose (Alopochen aegyptiacus)
R1 ☐	Spur-winged Goose (Plectropterus gambensis)
R4 ☐	African Pygmy Goose (Nettapus auritus)
V ☐	Mallard (Anas platyrhynchos)
Pw4 ☐	Common Teal (Anas crecca)
(Pw)4 ☐	Northern Pintail (Anas acuta)
Pw3 ☐	Garganey (Anas querquedula)
(Pw)4 ☐	Northern Shoveler (Anas clypeata)
V ☐	Common Pochard (Aythya ferina)
V ☐	Ferruginous Duck (Aythya nyroca)
V ☐	Tufted Duck (Aythya fuligula)
Ad1 ☐	Knob-billed Duck (Sarkidiornis melanotos)
V ☐	African Cuckoo-Falcon (Aviceda cuculoides)
(Ppw)4 ☐	European Honey-Buzzard (Pernis apivorus)
(?) ☐	Bat Hawk (Machaerhamphus alcinus)

Ad/(R)(2) ☐	Black-shouldered Kite (Elanus caeruleus)
Ad4 ☐	African Swallow-tailed Kite (Chelictinia riocourii)
V ☐	Red Kite (Milvus milvus)
Pw1 ☐	Black Kite (Milvus migrans migrans)
AdB1 ☐	'Yellow-billed' Kite (Milvus migrans parasitus)
R2 ☐	African Fish Eagle (Haliaeetus vocifer)
R1 ☐	Palm-nut Vulture (Gypohierax angolensis)
V ☐	Egyptian Vulture (Neophron percnopterus)
R1 ☐	Hooded Vulture (Necrosyrtes monachus)
R1 ☐	White-backed Vulture (Gyps africanus)
R2 ☐	Rüppell's Vulture (Gyps rueppellii)
V ☐	Lappet-faced Vulture (Torgos tracheliotus)
R4 ☐	White-headed Vulture (Trigonoceps occipitalis)
Ppw4 ☐	Short-toed Eagle (Circaetus gallicus gallicus)
R4 ☐	'Beaudouin's' Short-toed Eagle (Circaetus gallicus beaudouini)
R3 ☐	Brown Snake Eagle (Circaetus cinereus)
R4 ☐	Western Banded Snake Eagle (Circaetus cinerascens)
R4 ☐	Bateleur (Terathopius ecaudatus)
R1 ☐	African Harrier-Hawk (Polyboroides typus)
Ppw2 ☐	Western Marsh Harrier (Circus aeruginosus)
Ppw4 ☐	Pallid Harrier (Circus macrourus)
Pw4 ☐	Montagu's Harrier (Circus pygargus)
R2 ☐	Dark Chanting Goshawk (Melierax metabates)
(R)4 ☐	Gabar Goshawk (Melierax gabar)
(?) ☐	Great Sparrowhawk (Accipiter melanoleucus)
V ☐	Eurasian Sparrowhawk (Accipiter nisus)
R4 ☐	Western Little Sparrowhawk (Accipiter erythropus)
R4 ☐	African Goshawk (Accipiter tachiro)
R1 ☐	Shikra (Accipiter badius)
Ad/(R)2 ☐	Grasshopper Buzzard (Butastur rufipennis)
R1 ☐	Lizard Buzzard (Kaupifalco monogrammicus)
V ☐	Common Buzzard (Buteo buteo)
V ☐	Long-legged Buzzard (Buteo rufinus)
(Ad)4 ☐	Red-necked Buzzard (Buteo auguralis)
R4 ☐	Wahlberg's Eagle (Aquila wahlbergi)
R4 ☐	Tawny Eagle (Aquila rapax)
R4 ☐	African Hawk Eagle (Hieraaetus spilogaster)
(Pw)4 ☐	Booted Eagle (Hieraaetus pennatus)
R2 ☐	Long-crested Eagle (Lophaetus occipitalis)
R4 ☐	Martial Eagle (Polemaetus bellicosus)
Pw2 ☐	Osprey (Pandion haliaetus)
V ☐	Secretary Bird (Sagittarius serpentarius)
(Pw)4 ☐	Lesser Kestrel (Falco naumanni)
Pw2 ☐	Common Kestrel (Falco tinnunculus)
V ☐	Fox Kestrel (Falco alopex)
R1 ☐	Grey Kestrel (Falco ardosiaceus)
R4 ☐	Red-necked Falcon (Falco chicquera)
Pp(w)4 ☐	Eurasian Hobby (Falco subbuteo)
R4 ☐	African Hobby (Falco cuvieri)
A-4 ☐	Lanner Falcon (Falco biarmicus)

(Ppw)4 ☐	Peregrine Falcon (Falco peregrinus)
(?) ☐	White-throated Francolin (Francolinus albogularis)
R1 ☐	Double-spurred Francolin (Francolinus bicalcaratus)
R4 ☐	Ahanta Francolin (Francolinus ahantensis)
(Pw)4 ☐	Common Quail (Coturnix coturnix)
R2 ☐	Stone Partridge (Ptilopachus petrosus)
R4 ☐	Helmeted Guineafowl (Numida meleagris)
R4 ☐	Small Buttonquail (Turnix sylvatica)
R4 ☐	White-spotted Flufftail (Sarothrura pulchra)
R4 ☐	African Crake (Crecopsis egregia)
R2 ☐	Black Crake (Amaurornis flavirostris)
V ☐	Spotted Crake (Porzana porzana)
(Ad)4 ☐	Purple Swamphen (Porphyrio porphyrio)
(R)4 ☐	Allen's Gallinule (Porphyrula alleni)
Pw/A-/(R)3 ☐	Common Moorhen (Gallinula chloropus)
(R)4 ☐	Lesser Moorhen (Gallinula angulata)
R3 ☐	Black Crowned-Crane (Balearica pavonina)
R4 ☐	African Finfoot (Podica senegalensis)
V ☐	Denham's Bustard (Neotis denhami)
V ☐	Arabian Bustard (Ardeotis arabs)
Vb* ☐	White-bellied Bustard (Eupodotis senegalensis)
R4 ☐	Black-bellied Bustard (Eupodotis melanogaster)
R2 ☐	African Jacana (Actophilornis africanus)
A-(b)4 ☐	Greater Painted-snipe (Rostratula benghalensis)
Ppw1 ☐	Eurasian Oystercatcher (Haematopus ostralegus)
Ppw1 ☐	Black-winged Stilt (Himantopus himantopus)
Ppw3 ☐	Pied Avocet (Recurvirostra avosetta)
R1 ☐	Senegal Thick-knee (Burhinus senegalensis)
(?) ☐	Spotted Thick-knee (Burhinus capensis)
Ad3 ☐	Egyptian Plover (Pluvianus aegyptius)
R4 ☐	Bronze-winged Courser (Rhinoptilus chalcopterus)
V ☐	Cream-coloured Courser (Cursorius cursor)
R3 ☐	Temminck's Courser (Cursorius temminckii)
Pw/(R)2 ☐	Collared Pratincole (Glareola pratincola)
Ppw4 ☐	Little Ringed Plover (Charadrius dubius)
Ppw1 ☐	Ringed Plover (Charadrius hiaticula)
Ad4 ☐	Kittlitz's Plover (Charadrius pecuarius)
V ☐	Forbes's Plover (Charadrius forbesi)
Ppw4 ☐	Kentish Plover (Charadrius alexandrinus)
R4 ☐	White-fronted Plover (Charadrius marginatus)
V ☐	Eurasian Dotterel (Eudromias morinellus)
V ☐	European Golden Plover (Pluvialis apricaria)
V ☐	American Golden Plover (Pluvialis dominica)
Ppw1 ☐	Grey Plover (Pluvialis squatarola)
R1 ☐	Wattled Plover (Vanellus senegallus)
(Ad)4 ☐	White-crowned Plover (Vanellus albiceps)
R1 ☐	Black-headed Plover (Vanellus tectus)
R1 ☐	Spur-winged Plover (Vanellus spinosus)
V ☐	Northern Lapwing (Vanellus vanellus)
Pw2 ☐	Common Snipe (Gallinago gallinago)

V ☐	Great Snipe (Gallinago media)
(Pw)4 ☐	Jack Snipe (Lymnocryptes minimus)
Ppw1 ☐	Black-tailed Godwit (Limosa limosa)
Ppw2 ☐	Bar-tailed Godwit (Limosa lapponica)
Ppw1 ☐	Whimbrel (Numenius phaeopus)
Ppw3 ☐	Eurasian Curlew (Numenius arquata)
Ppw4 ☐	Spotted Redshank (Tringa erythropus)
Ppw1 ☐	Common Redshank (Tringa totanus)
Ppw2 ☐	Marsh Sandpiper (Tringa stagnatilis)
Ppw1 ☐	Common Greenshank (Tringa nebularia)
V ☐	Lesser Yellowlegs (Tringa flavipes)
Ppw2 ☐	Green Sandpiper (Tringa ochropus)
Ppw1 ☐	Wood Sandpiper (Tringa glareola)
V ☐	Terek Sandpiper (Xenus cinereus)
Ppw1 ☐	Common Sandpiper (Actitis hypoleucos)
Ppw1 ☐	Ruddy Turnstone (Arenaria interpres)
Ppw3 ☐	Red Knot (Calidris canutus)
Ppw1 ☐	Sanderling (Calidris alba)
Ppw1 ☐	Little Stint (Calidris minuta)
(Ppw)4 ☐	Temminck's Stint (Calidris temminckii)
Ppw1 ☐	Curlew Sandpiper (Calidris ferruginea)
Ppw2 ☐	Dunlin (Calidris alpina)
Ppw1 ☐	Ruff (Philomachus pugnax)
(Pp)4 ☐	Grey Phalarope (Phalaropus fulicarius)
Pp4 ☐	Pomarine Skua (Stercorarius pomarinus)
Pp4 ☐	Arctic Skua (Stercorarius parasiticus)
V ☐	Audouin's Gull (Larus audouinii)
V ☐	Common Gull (Larus canus)
Pw1 ☐	Lesser Black-backed Gull (Larus fuscus)
Pw4 ☐	Yellow-legged Herring Gull (Larus cachinnans)
V ☐	Mediterranean Gull (Larus melanocephalus)
V ☐	Laughing Gull (Larus atricilla)
V ☐	Franklin's Gull (Larus pipixcan)
V ☐	Little Gull (Larus minutus)
Pw3 ☐	Black-headed Gull (Larus ridibundus)
R1 ☐	Grey-headed Gull (Larus cirrocephalus)
Pw3 ☐	Slender-billed Gull (Larus genei)
V ☐	Black-legged Kittiwake (Rissa tridactyla)
Pw4 ☐	Gull-billed Tern (Sterna nilotica)
R1 ☐	Caspian Tern (Sterna caspia)
R1 ☐	Royal Tern (Sterna maxima)
Pw4 ☐	Lesser Crested Tern (Sterna bengalensis)
Ppw1 ☐	Sandwich Tern (Sterna sandvicensis)
V/(Pp)4 ☐	Roseate Tern (Sterna dougallii)
Pp3 ☐	Common Tern (Sterna hirundo)
Pp4 ☐	Arctic Tern (Sterna paradisaea)
Pp/(R)2 ☐	Little Tern (Sterna albifrons)
Ppw4 ☐	Whiskered Tern (Chlidonias hybridus)
Ppw1 ☐	Black Tern (Chlidonias niger)
Pp1 ☐	White-winged Black Tern (Chlidonias leucopterus)

V ☐	Brown Noddy (Anous stolidus)
V ☐	Black Noddy (Anous minutus)
A-4 ☐	African Skimmer (Rynchops flavirostris)
(V) ☐	Chestnut-bellied Sandgrouse (Pterocles exustus)
R3 ☐	Four-banded Sandgrouse (Pterocles quadricinctus)
R1 ☐	Speckled Pigeon (Columba guinea)
Pp(4) ☐	European Turtle Dove (Streptopelia turtur)
R1 ☐	Laughing Dove (Streptopelia senegalensis)
R1 ☐	African Mourning Dove (Streptopelia decipiens)
R1 ☐	Vinaceous Dove (Streptopelia vinacea)
(R)4 ☐	African Collared Dove (Streptopelia roseogrisea)
R1 ☐	Red-eyed Dove (Streptopelia semitorquata)
R1 ☐	Black-billed Wood Dove (Turtur abyssinicus)
R1 ☐	Blue-spotted Wood Dove (Turtur afer)
Ad1 ☐	Namaqua Dove (Oena capensis)
R2 ☐	Bruce's Green Pigeon (Treron waalia)
R2 ☐	African Green Pigeon (Treron calva)
R4 ☐	Brown-necked Parrot (Poicephalus robustus)
R1 ☐	Senegal Parrot (Poicephalus senegalus)
R1 ☐	Rose-ringed Parakeet (Psittacula krameri)
R4 ☐	Guinea Turaco (Tauraco persa)
R3 ☐	Violet Turaco (Musophaga violacea)
R1 ☐	Western Grey Plantain-eater (Crinifer piscator)
R/A-/Pw4 ☐	Great Spotted Cuckoo (Clamator glandarius)
(R)4 ☐	Jacobin Cuckoo (Clamator jacobinus)
A-B1 ☐	Levaillant's Cuckoo (Oxylophus levaillantii)
V ☐	Red-chested Cuckoo (Cuculus solitarius)
V ☐	Black Cuckoo (Cuculus clamosus)
Pw3 ☐	Common Cuckoo (Cuculus canorus)
A-B3 ☐	African Cuckoo (Cuculus gularis)
V(b*) ☐	African Emerald Cuckoo (Chrysococcyx cupreus)
Ad2 ☐	Klaas's Cuckoo (Chrysococcyx klaas)
A-B2 ☐	Didric Cuckoo (Chrysococcyx caprius)
(?) ☐	Yellowbill (Ceuthmochares aereus)
R4 ☐	Black Coucal (Centropus grillii)
R1 ☐	Senegal Coucal (Centropus senegalensis)
R2 ☐	Barn Owl (Tyto alba)
R3 ☐	African Scops Owl (Otus senegalensis)
R2 ☐	White-faced Scops Owl (Otus leucotis)
(R)4 ☐	Spotted Eagle Owl (Bubo africanus)
R3 ☐	Verreaux's Eagle Owl (Bubo lacteus)
R4 ☐	Pel's Fishing-Owl (Scotopelia peli)
R2 ☐	Pearl-spotted Owlet (Glaucidium perlatum)
(Aw)4 ☐	Marsh Owl (Asio capensis)
Pw4 ☐	European Nightjar (Caprimulgus europaeus)
V ☐	Swamp Nightjar (Caprimulgus natalensis)
(A-b)3 ☐	Long-tailed Nightjar (Caprimulgus climacurus)
R/(A-b)2 ☐	Standard-winged Nightjar (Macrodipteryx longipennis)
V ☐	Pennant-winged Nightjar (Macrodipteryx vexillarius)
R4 ☐	Mottled Spinetail (Telacanthura ussheri)

R1 ☐	African Palm Swift (Cypsiurus parvus)
Ppw3 ☐	Pallid Swift (Apus pallidus)
Pp1 ☐	Common Swift (Apus apus)
R1 ☐	Little Swift (Apus affinis)
Vb* ☐	White-rumped Swift (Apus caffer)
V ☐	Blue-naped Mousebird (Urocolius macrourus)
A-4 ☐	Shining-blue Kingfisher (Alcedo quadribrachys)
R2 ☐	Malachite Kingfisher (Alcedo cristata)
R/AwB3 ☐	African Pygmy Kingfisher (Ceyx pictus)
R2 ☐	Grey-headed Kingfisher (Halcyon leucocephala)
R3 ☐	Blue-breasted Kingfisher (Halcyon malimbica)
A-B1 ☐	Woodland Kingfisher (Halcyon senegalensis)
R2 ☐	Striped Kingfisher (Halcyon chelicuti)
R3 ☐	Giant Kingfisher (Megaceryle maxima)
R1 ☐	Pied Kingfisher (Ceryle rudis)
R1 ☐	Little Bee-eater (Merops pusillus)
R3 ☐	Swallow-tailed Bee-eater (Merops hirundineus)
Aw/(R)2 ☐	Red-throated Bee-eater (Merops bullocki)
A-(2) ☐	White-throated Bee-eater (Merops albicollis)
Ad4 ☐	Little Green Bee-eater (Merops orientalis)
Ppw1 ☐	Blue-cheeked Bee-eater (Merops persicus)
Pp(w)4 ☐	European Bee-eater (Merops apiaster)
Ad/(R)1 ☐	Northern Carmine Bee-eater (Merops nubicus)
(Ppw)4 ☐	European Roller (Coracias garrulus)
Ad/R1 ☐	Abyssinian Roller (Coracias abyssinica)
R2 ☐	Rufous-crowned Roller (Coracias naevia)
R1 ☐	Blue-bellied Roller (Coracias cyanogaster)
R1 ☐	Broad-billed Roller (Eurystomus glaucurus)
R1 ☐	Green Wood-Hoopoe (Phoeniculus purpureus)
R4 ☐	Black Scimitarbill (Rhinopomastus aterrimus)
R/(Pw)4 ☐	Eurasian Hoopoe (Upupa epops)
R1 ☐	Red-billed Hornbill (Tockus erythrorhynchus)
R3 ☐	African Pied Hornbill (Tockus fasciatus)
R1 ☐	African Grey Hornbill (Tockus nasutus)
R3 ☐	Abyssinian Ground Hornbill (Bucorvus abyssinicus)
R1 ☐	Yellow-fronted Tinkerbird (Pogoniulus chrysoconus)
R4 ☐	Yellow-rumped Tinkerbird (Pogoniulus bilineatus)
R3 ☐	Vieillot's Barbet (Lybius vieilloti)
R1 ☐	Bearded Barbet (Lybius dubius)
R4 ☐	Spotted Honeyguide (Indicator maculatus)
R2 ☐	Greater Honeyguide (Indicator indicator)
R3 ☐	Lesser Honeyguide (Indicator minor)
Ppw4 ☐	Eurasian Wryneck (Jynx torquilla)
R1 ☐	Fine-spotted Woodpecker (Campethera punctuligera)
R4 ☐	Golden-tailed Woodpecker (Campethera abingoni)
R4 ☐	Buff-spotted Woodpecker (Campethera nivosa)
(?) ☐	Little Grey Woodpecker (Dendropicos elachus)
R4 ☐	Cardinal Woodpecker (Dendropicos fuscescens)
R1 ☐	Grey Woodpecker (Dendropicos goertae)
R4 ☐	Brown-backed Woodpecker (Picoides obsoletus)

V ☐	Singing Bush-lark (Mirafra cantillans)
(R)4 ☐	Flappet Lark (Mirafra rufocinnamomea)
V ☐	Rufous-rumped Lark (Pinarocorys erythropygia)
R1 ☐	Crested Lark (Galerida cristata)
R3 ☐	Chestnut-backed Sparrow-Lark (Eremopterix leucotis)
Ad/(R)2 ☐	Fanti Saw-wing (Psalidoprocne obscura)
Ppw1 ☐	Sand Martin (Riparia riparia)
V ☐	Brown-throated Sand Martin (Riparia paludicola)
V(b*) ☐	Banded Martin (Riparia cincta)
V ☐	Grey-rumped Swallow (Pseudohirundo griseopyga)
R4 ☐	Rufous-chested Swallow (Hirundo semirufa)
A-B/R(1) ☐	Mosque Swallow (Hirundo senegalensis)
V ☐	Lesser Striped Swallow (Hirundo abyssinica)
R2 ☐	Red-rumped Swallow (Hirundo daurica)
R1 ☐	Wire-tailed Swallow (Hirundo smithii)
R3 ☐	Pied-winged Swallow (Hirundo leucosoma)
Pp1 ☐	Barn Swallow (Hirundo rustica)
R1 ☐	Red-chested Swallow (Hirundo lucida)
Ppw3 ☐	House Martin (Delichon urbica)
Ppw(1) ☐	Yellow Wagtail (Motacilla flava)
(Pw)4 ☐	Grey Wagtail (Motacilla cinerea)
Pw1 ☐	White Wagtail (Motacilla alba)
V ☐	African Pied Wagtail (Motacilla aguimp)
Pw4 ☐	Tawny Pipit (Anthus campestris)
R3 ☐	Plain-backed Pipit (Anthus leucophrys)
Ppw4 ☐	Tree Pipit (Anthus trivialis)
Ppw4 ☐	Red-throated Pipit (Anthus cervinus)
R3 ☐	Yellow-throated Longclaw (Macronyx croceus)
R4 ☐	Red-shouldered Cuckoo-shrike (Campephaga phoenicea)
R4 ☐	White-breasted Cuckoo-shrike (Coracina pectoralis)
R2 ☐	Little Greenbul (Andropadus virens)
R2 ☐	Yellow-throated Leaflove (Chlorocichla flavicollis)
V ☐	Swamp Palm Greenbul (Thescelocichla leucopleura)
R4 ☐	Leaf-love (Pyrrhurus scandens)
R4 ☐	Grey-headed Bristlebill (Bleda canicapilla)
R1 ☐	Common Bulbul (Pycnonotus barbatus)
(Pw)4 ☐	Blue Rock Thrush (Monticola solitarius)
(Pw)4 ☐	European Rock Thrush (Monticola saxatilis)
R1 ☐	African Thrush (Turdus pelios)
Ppw2 ☐	Nightingale (Luscinia megarhynchos)
V ☐	Bluethroat (Luscinia svecica)
R3 ☐	Snowy-crowned Robin-Chat (Cossypha niveicapilla)
R2 ☐	White-crowned Robin-Chat (Cossypha albicapilla)
V ☐	Rufous Scrub-Robin (Cercotrichas galactotes)
Ppw2 ☐	Common Redstart (Phoenicurus phoenicurus)
Ppw1 ☐	Whinchat (Saxicola rubetra)
Ppw2 ☐	Northern Wheatear (Oenanthe oenanthe)
V ☐	Black-eared Wheatear (Oenanthe hispanica)
V ☐	Isabelline Wheatear (Oenanthe isabellina)
V ☐	Blackstart (Cercomela melanura)

Full species list

Ad/(R)3 ☐	Northern Anteater-Chat (Myrmecocichla aethiops)
R4 ☐	White-fronted Black Chat (Myrmecocichla albifrons)
V ☐	Grasshopper Warbler (Locustella naevia)
Ppw3 ☐	Sedge Warbler (Acrocephalus schoenobaenus)
Ppw3 ☐	Reed Warbler (Acrocephalus scirpaceus scirpaceus)
(R)4 ☐	'African' Reed Warbler (Acrocephalus s. baeticatus)
V ☐	Great Reed Warbler (Acrocephalus arundinaceus)
Ppw2 ☐	Olivaceous Warbler (Hippolais pallida)
Ppw1 ☐	Melodious Warbler (Hippolais polyglotta)
R1 ☐	Green-backed Eremomela (Eremomela pusilla)
(R4) ☐	Green Crombec (Sylvietta virens)
V ☐	Lemon-bellied Crombec (Sylvietta denti)
R3 ☐	Northern Crombec (Sylvietta brachyura)
Ppw1 ☐	Willow Warbler (Phylloscopus trochilus)
Ppw1 ☐	Common Chiffchaff (Phylloscopus collybita)
Ppw4 ☐	Bonelli's Warbler (Phylloscopus bonelli)
V ☐	Wood Warbler (Phylloscopus sibilatrix)
R4 ☐	Yellow-bellied Hyliota (Hyliota flavigaster)
R4 ☐	Green Hylia (Hylia prasina)
V ☐	Orphean Warbler (Sylvia hortensis)
Ppw2 ☐	Garden Warbler (Sylvia borin)
Ppw(2) ☐	Blackcap (Sylvia atricapilla)
Ppw4 ☐	Common Whitethroat (Sylvia communis)
Pw4 ☐	Subalpine Warbler (Sylvia cantillans)
R1 ☐	Zitting Cisticola (Cisticola juncidis)
R4 ☐	Croaking Cisticola (Cisticola natalensis)
(?) ☐	Red-pate Cisticola (Cisticola ruficeps)
(?) ☐	Rufous Cisticola (Cisticola rufus)
(R/AwB)2 ☐	Siffling Cisticola (Cisticola brachypterus)
R4 ☐	Whistling Cisticola (Cisticola lateralis)
R2 ☐	Red-faced Cisticola (Cisticola erythrops)
R2 ☐	Singing Cisticola (Cisticola cantans)
R2 ☐	Winding Cisticola (Cisticola galactotes)
R1 ☐	Tawny-flanked Prinia (Prinia subflava)
R4 ☐	Red-winged Warbler (Heliolais erythroptera)
R4 ☐	Yellow-breasted Apalis (Apalis flavida)
R1 ☐	Grey-backed Camaroptera (Camaroptera brachyura)
R3 ☐	Oriole Warbler (Hypergerus atriceps)
R4 ☐	Pale Flycatcher (Bradornis pallidus)
R2 ☐	Northern Black-Flycatcher (Melaenornis edolioides)
Ppw2 ☐	Pied Flycatcher (Ficedula hypoleuca)
Ppw2 ☐	Spotted Flycatcher (Muscicapa striata)
R2 ☐	Swamp Flycatcher (Muscicapa aquatica)
R4 ☐	Lead-coloured Flycatcher (Myioparus plumbeus)
V ☐	Shrike-Flycatcher (Megabyas flammulatus)
V ☐	Black-and-white Flycatcher (Bias musicus)
R3 ☐	Senegal Batis (Batis senegalensis)
R2 ☐	Brown-throated Wattle-eye (Platysteira cyanea)
(R)4 ☐	Blue Flycatcher (Elminia longicaudata)
R2 ☐	African Paradise-Flycatcher (Terpsiphone viridis)*

Full species list

R2 ☐	Red-bellied Paradise-Flycatcher (Terpsiphone rufiventer)*
R1 ☐	Brown Babbler (Turdoides plebejus)
R1 ☐	Blackcap Babbler (Turdoides reinwardii)
(R4) ☐	Capuchin Babbler (Phyllanthus atripennis)
R4 ☐	White-winged Black Tit (Parus leucomelas)
R4 ☐	Yellow Penduline-Tit (Remiz parvulus)
V ☐	Spotted Creeper (Salpornis spilonotus)
R4 ☐	Mouse-brown Sunbird (Anthreptes gabonicus)
R4 ☐	Western Violet-backed Sunbird (Anthreptes longuemarei)
R4 ☐	Collared Sunbird (Anthreptes collaris)
R4 ☐	Pygmy Sunbird (Anthreptes platurus)
R4 ☐	Green-headed Sunbird (Nectarinia verticalis)
R2 ☐	Scarlet-chested Sunbird (Nectarinia senegalensis)
R1 ☐	Variable Sunbird (Nectarinia venusta)
R2 ☐	Copper Sunbird (Nectarinia cuprea)
R1 ☐	Splendid Sunbird (Nectarinia coccinigaster)
R1 ☐	Beautiful Sunbird (Nectarinia pulchella)
R4 ☐	African Yellow White-eye (Zosterops senegalensis)
(Pw4) ☐	European Golden Oriole (Oriolus oriolus)
R2 ☐	African Golden Oriole (Oriolus auratus)
V ☐	Red-backed Shrike (Lanius collurio)
V ☐	Great Grey Shrike (Lanius excubitor)
Ppw1 ☐	Woodchat Shrike (Lanius senator)
R1 ☐	Yellow-billed Shrike (Corvinella corvina)
(R/A-b)4 ☐	Brubru (Nilaus afer)
R2 ☐	Northern Puffback (Dryoscopus gambensis)
R1 ☐	Black-crowned Tchagra (Tchagra senegala)
R1 ☐	Yellow-crowned Gonolek (Laniarius barbarus)
R4 ☐	Sulphur-breasted Bushshrike (Malaconotus sulfureopectus)
R3 ☐	Grey-headed Bushshrike (Malaconotus blanchoti)
(?) ☐	Western Nicator (Nicator chloris)
R3 ☐	White-crested Helmet-Shrike (Prionops plumatus)
(R4) ☐	Square-tailed Drongo (Dicrurus ludwigii)
R1 ☐	Fork-tailed Drongo (Dicrurus adsimilis)
R1 ☐	Piapiac (Ptilostomus afer)
R1 ☐	Pied Crow (Corvus albus)
V ☐	Brown-necked Raven (Corvus ruficollis)
R1 ☐	Purple Glossy Starling (Lamprotornis purpureus)
R(2) ☐	Bronze-tailed Glossy Starling (Lamprotornis chalcurus)
R1 ☐	Greater Blue-eared Glossy Starling (Lamprotornis chalybaeus)
(R)3 ☐	Lesser Blue-eared Glossy Starling (Lamprotornis chloropterus)
Ad4 ☐	Splendid Glossy Starling (Lamprotornis splendidus)
R1 ☐	Long-tailed Glossy Starling (Lamprotornis caudatus)
R2 ☐	Chestnut-bellied Starling (Lamprotornis pulcher)
A-(b)4 ☐	Violet-backed Starling (Cinnyricinclus leucogaster)
V ☐	Wattled Starling (Creatophora cinerea)
R2 ☐	Yellow-billed Oxpecker (Buphagus africanus)
R1 ☐	House Sparrow (Passer domesticus)
R1 ☐	Grey-headed Sparrow (Passer griseus)
(Ad)4 ☐	Sudan Golden Sparrow (Passer luteus)

Full species list

R1 ☐	Bush Petronia (Petronia dentata)
R1 ☐	White-billed Buffalo Weaver (Bubalornis albirostris)
Ad4 ☐	Speckle-fronted Weaver (Sporopipes frontalis)
R4 ☐	Chestnut-crowned Sparrow-Weaver (Plocepasser superciliosus)
R2 ☐	Little Weaver (Ploceus luteolus)
R3 ☐	Black-necked Weaver (Ploceus nigricollis)
R4 ☐	Southern Masked Weaver (Ploceus velatus)
R4 ☐	Heuglin's Masked Weaver (Ploceus heuglini)
(R)4 ☐	Vieillot's Black Weaver (Ploceus nigerrimus)
R1 ☐	Village Weaver (Ploceus cucullatus)
R2 ☐	Black-headed Weaver (Ploceus melanocephalus)
V ☐	Red-headed Weaver (Anaplectes rubriceps)
R4 ☐	Red-headed Quelea (Quelea erythrops)
(Ad)4 ☐	Red-billed Quelea (Quelea quelea)
R2 ☐	Yellow-crowned Bishop (Euplectes afer)
R2 ☐	Black-winged Red Bishop (Euplectes hordeaceus)
R1 ☐	Northern Red Bishop (Euplectes franciscanus)
R2 ☐	Yellow-shouldered Widowbird (Euplectes macrourus)
V ☐	Red-collared Widowbird (Euplectes ardens)
V ☐	Parasitic Weaver (Anomalospiza imberbis)
V ☐	Chestnut-breasted Negro-Finch (Nigrita bicolor)
V ☐	Grey-headed Olive-back (Nesocharis capistrata)
V ☐	Green-winged Pytilia (Pytilia melba)
R4 ☐	Red-winged Pytilia (Pytilia phoenicoptera)
R4 ☐	Crimson Seed-cracker (Pyrenestes sanguineus)
R4 ☐	Western Bluebill (Spermophaga haematina)
R4 ☐	Bar-breasted Firefinch (Lagonosticta rufopicta)
R1 ☐	Red-billed Firefinch (Lagonosticta senegala)
R4 ☐	Black-faced Firefinch (Lagonosticta larvata)
R2 ☐	Lavender Waxbill (Estrilda caerulescens)
R1 ☐	Orange-cheeked Waxbill (Estrilda melpoda)
R2 ☐	Black-rumped Waxbill (Estrilda troglodytes)
R1 ☐	Red-cheeked Cordon-bleu (Uraeginthus bengalus)
(Ad)4 ☐	Zebra Waxbill (Amandava subflava)
R2 ☐	Quailfinch (Ortygospiza atricollis)
R4 ☐	African Silverbill (Lonchura cantans)
R1 ☐	Bronze Mannikin (Lonchura cucullata)
(?) ☐	Magpie Mannikin (Lonchura fringilloides)
Ad4 ☐	Cut-throat (Amadina fasciata)
R1 ☐	Village Indigobird (Vidua chalybeata)
R4 ☐	Baka Indigobird (Vidua larvaticola)
R3 ☐	Pin-tailed Whydah (Vidua macroura)
R4 ☐	Long-tailed Paradise Whydah (Vidua paradisaea)
R2 ☐	White-rumped Seedeater (Serinus leucopygius)
R1 ☐	Yellow-fronted Canary (Serinus mozambicus)
V ☐	Ortolan Bunting (Emberiza hortulana)
V ☐	House Bunting (Emberiza striolata)
AdB3 ☐	Cinnamon-breasted Bunting (Emberiza tahapisi)
Awb4 ☐	Brown-rumped Bunting (Emberiza affinis)

* Red-bellied Paradise Flycatcher and Paradise Flycatcher. Gore (1990) suggests that these are probably conspecific since they often occur together and court indiscriminately. I have not distinguished the two in the site descriptions where "Paradise Flycatcher" may refer to either.

ALTERNATIVE VERNACULAR NAMES

The existence and general usage of more than one vernacular name for the same bird always causes confusion, and nowhere more so than in The Gambia. Below is a list of those most likely to be encountered.

Great Cormorant = White-breasted Cormorant
Long-tailed Cormorant = Long-tailed Shag or Reed Cormorant
Pink-backed Pelican = Grey Pelican
African Darter = Snake Bird
Black Egret = Black Heron
Western Reef-Egret = Reef Heron
Intermediate Egret = Yellow-billed Egret
Yellow-billed Stork = Wood Ibis
Woolly-necked Stork = White-necked Stork
White-faced Whistling -Duck = White-faced Tree-Duck
Knob-billed Duck = Knob-billed Goose
African Fish Eagle = African River Eagle
Short-toed Eagle = Snake Eagle
Brown Snake Eagle = Brown Harrier Eagle
Western Banded Snake Eagle = Banded Harrier Eagle
Great Sparrowhawk = Black Sparrowhawk
Common Buzzard = Steppe Buzzard
Red-necked Buzzard = Red-tailed or Augur Buzzard
Long-crested Eagle = Long-crested Hawk-eagle
Red-necked Falcon = Red-necked Kestrel
Small Buttonquail = Andalusian Hemipode or African Button-Quail
White-spotted Flufftail = White-spotted Pigmy Rail or Pygmy Crake
Purple Swamphen = Purple Gallinule or King Reed-hen
Arabian Bustard = Sudan Bustard
White-bellied Bustard = Senegal Bustard
African Jacana = Lily-trotter
Bronze-winged Courser = Violet-tipped Courser
Egyptian Plover = Crocodile Bird
Wattled Plover = Senegal Wattled Plover
White-crowned Plover = White-headed Plover
African Collared Dove = Rose-grey Dove or Pink-headed Dove
Blue-spotted Wood Dove = Red-billed Wood Dove
Namaqua Dove = Long-tailed Dove

Alternative vernacular names

Bruce's Green Pigeon = Yellow-bellied Fruit-Pigeon
African Green Pigeon = Green Fruit-Pigeon
Senegal Parrot = Yellow-bellied Parrot
Rose-ringed Parakeet = Senegal Long-tailed or Ring-necked
                              Parakeet
Guinea Turaco = Green-crested Touraco
Violet Turaco = Violet Plantain-eater
Jacobin Cuckoo = Pied Crested Cuckoo
Yellowbill = Yellowbill Coucal
White-faced Scops Owl = White-faced Owl
Verreaux's Eagle-Owl = Milky Eagle-Owl
Swamp Nightjar = White-tailed Nightjar
Mottled Spinetail = Mottled-throated Spinetail or
                         Ussher's Spine-tailed Swift
Woodland Kingfisher = Senegal Kingfisher
Green Wood-Hoopoe = Senegal Wood-Hoopoe
Black Scimitarbill = Black or Lesser Wood-Hoopoe
Red-billed Hornbill = Red-beaked Hornbill
African Pied Hornbill = Allied or Black-and-white-tailed Hornbill
Yellow-fronted Tinkerbird = Yellow-fronted Barbet
Yellow-rumped Tinkerbird = Lemon-rumped Tinkerbird
Greater Honeyguide = Black-throated Honeyguide
Little Grey Woodpecker = Least Grey Woodpecker
Brown-backed Woodpecker = Lesser White-spotted Woodpecker
Chestnut-backed Sparrow-Lark = Chestnut-backed Finch-Lark
Fanti Saw-wing = Fanti Rough-winged Swallow
Brown-throated Sand Martin = African Sand Martin
Little Greenbul = Little Green Bulbul
Common Bulbul = Common Garden Bulbul
African Thrush = Kurrichane or Olive Thrush
Rufous Scrub-Robin = Rufous Bush-Chat
Blackstart = Black-tailed Rock-Chat
Northern Anteater-Chat = Ant-Chat
Northern Crombec = Nuthatch Warbler
Yellow-bellied Hyliota = Yellow-bellied Flycatcher
Zitting Cisticola = Fan-tailed Warbler
Croaking Cisticola = Striped Cisticola
Siffling Cisticola = Shortwing Cisticola
Winding Cisticola = Rufous Grass-Warbler
Tawny-flanked Prinia = West African Prinia
Oriole Warbler = Moho
Lead-coloured Flycatcher = Grey Tit-Babbler
Senegal Batis = Senegal Puff-back Flycatcher
Brown-throated Wattle-eye = Scarlet-spectacled Wattle-eye
Blue Flycatcher = Blue Fairy Flycatcher
White-winged Black Tit = White-shouldered Black Tit
Yellow Penduline-Tit = West African Penduline Tit
Pygmy Sunbird = Pigmy Long-tailed Sunbird
Green-headed Sunbird = Olive-backed Sunbird
Variable Sunbird = Yellow-bellied Sunbird

Beautiful Sunbird = Beautiful Long-tailed Sunbird
Yellow-billed Shrike = Long-tailed Shrike
Brubru = Brubru Shrike
Northern Puffback = Gambian Puff-back Shrike
Black-crowned Tchagra = Black-headed Bush-Shrike
Yellow-crowned Gonolek = Barbary Shrike
Sulphur-breasted Bushshrike = Orange-breasted Bush-Shrike
Grey-headed Bushshrike = Gladiator Bush-Shrike
White-crested Helmet-Shrike = Long-crested Helmet-Shrike
Fork-tailed Drongo = Glossy-backed Drongo
Piapiac = Black Magpie
Bronze-tailed Glossy Starling = Short-tailed Glossy Starling
Violet-backed Starling = Amethyst or Plum-coloured Starling
Bush Petronia = Bush Sparrow
Speckle-fronted Weaver = Scaly-fronted Weaver
Black-necked Weaver = Spectacled Weaver
Southern Masked Weaver = Vitelline Masked Weaver
Red-headed Weaver = Red-winged Malimbe
Red-headed Quelea = Red-headed Dioch
Red-billed Quelea = Black-faced Dioch
Black-winged Red Bishop = Fire-crowned Bishop
Yellow-shouldered Widowbird = Yellow-mantled Whydah
Red-collared Widowbird = Long-tailed Black Whydah
Grey-headed Olive-back = White-cheeked Olive Weaver
Green-winged Pytilia = Melba Finch
Crimson Seed-cracker = Black-bellied Seed-cracker
Western Bluebill = Blue-billed Weaver
Red-billed Firefinch = Senegal Fire-Finch
Lavender Waxbill = Lavender Fire-Finch
Zebra Waxbill = Orange-breasted Waxbill
African Silverbill = Warbling Silverbill
Cut-throat = Cut-throat Weaver
Village Indigobird = Green Indigo-bird or Senegal Indigo Finch
Baka Indigobird = Cameroon Indigo Finch
Long-tailed Paradise Whydah = Broad-tailed Paradise Whydah
White-rumped Seedeater = Grey Canary
Cinnamon-breasted Bunting = Cinnamon-breasted Rock-Bunting
Brown-rumped Bunting = Nigerian Little Bunting

BUTTERFLIES

Selected list of commoner species compiled mainly from information supplied by Dr Torben B Larsen.

- [ ] Citrus Swallowtail (Papilio demodocus)
- [ ] African Emigrant (Catopsilia florella)
- [ ] Common Grass Yellow (Eurema hecabe)
- [ ] Fig Blue (Myrina silenus)
- [ ] African Monarch (Danaus chrysippus)
- [ ] Diadem (Hypolimnas misippus)
- [ ] Pearl Charaxes (Charaxes varanes)
- [ ] Painted Lady (Vanessa cardui)
- [ ] Yellow Pansy (Junonia hierta)
- [ ] Black Pansy (Junonia oenone)
- [ ] Blue Pansy (Junonia orithyia)
- [ ] Guineafowl (Hamanumida daedalus)
- [ ] Evening Brown (Melanitis leda)
- [ ] Senegal Policeman (Pyrrhiades aeschylus)
- [ ] Dark Flat (Sarangesa laelius)

AMPHIBIANS & REPTILES

Derived mainly from Andersson (1937) and Hakansson (1981). Some species have no traceable English vernacular name, in which cases the basic type of animal is given and marked with an asterisk.

- [ ] African Square-marked Toad (Bufo regularis)
- [ ] Mottled Burrowing Frog (Hemisus marmoratus sudanensis)
- [ ] Galam River Frog (Hylarana galamensis)
- [ ] Running Frog or Kassina* (Kassina cassinoides)
- [ ] Bubbling Kassina (Kassina senegalensis)
- [ ] Sharp-nosed or Ridged Frog* (Ptychadena aequiplicata)
- [ ] Maccarthy Island Ridged Frog (Ptychadena maccarthyensis)
- [ ] Sharp-nosed or Ridged Frog* (Ptychadena trinodis)
- [ ] African Bullfrog (Rana occipitalis)
- [ ] Spanish Terrapin (Clemmys caspica leprosa)
- [ ] Bell's Hinged Tortoise (Kinixys belliana)
- [ ] West African Hinge-backed Tortoise (Kinixys erosa)
- [ ] Soft-shelled River Turtle (Trionyx triunguis)
- [ ] Senegal Soft-shelled Turtle (Cyclanorbis senegalensis)
- [ ] Pan-hinged Terrapin (Pelusios subniger)
- [ ] Nile Crocodile (Crocodylus niloticus)
- [ ] Long-snouted Crocodile (Crocodylus cataphractus)
- [ ] Pigmy Crocodile (Osteolaemus tetraspis)
- [ ] Gecko* (Tarentola ephippiata)
- [ ] Dwarf Gecko* (Lygodactylus gutturalis)
- [ ] Brook's Gecko (Hemidactylus brooki angulatus)
- [ ] Common Agama (Agama agama)

☐ Chamaeleon* (Chamaeleo gracilis)
☐ Senegal Chamaeleon (Chamaeleo senegalensis)
☐ Armitage's Skink (Chalcides armitagei)
☐ Skink* (Mabuya affinis)
☐ Perrotet's Skink (Mabuya perrotetii)
☐ Nile Monitor (Varanus niloticus)
☐ Bosc's Monitor (Varanus exanthematicus)
☐ Spotted Blind Snake (Typhlops punctatus)
☐ Rock Python (Python sebae)
☐ Royal Python (Python regius)
☐ Common House Snake (Boaedon fuliginosum)
☐ West African House Snake (Boaedon virgatum)
☐ Smyth's Water Snake (Graya smythii)
☐ Wolf Snake* (Lycophidion semicinctum)
☐ Northern Green Snake (Philothamnus irregularis)
☐ Spotted Bush Snake (Philothamnus semivariegatus)
☐ Emerald Snake (Gastropyxis smaragdina)
☐ Shovel-snout Snake* (Prosymna meleagris)
☐ Egg-eating Snake* (Dasypeltis fasciata)
☐ Common Egg-eater (Dasypeltis scabra)
☐ Herald Snake (Crotaphopeltis hotamboeia)
☐ Striped Beauty Snake (Psammophis elegans)
☐ African Beauty Snake (Psammophis sibilans)
☐ Olive Grass Snake (Psammophis phillipsi)
☐ Sundevall's Garter Snake (Elapsoidea sundevalli)
☐ Black Cobra (Naja melanoleuca)
☐ Spitting Cobra (Naja nigricollis)
☐ Green Mamba (Dendroaspis viridis)
☐ Puff Adder (Bitis arietans)
☐ West African Night Adder (Causus maculatus)
☐ Common Night Adder (Causus rhombeatus)
☐ Saw-scaled Viper (Echis carinatus pyramidium)

LARGER MAMMALS

This list is compiled from various sources. Although all the species listed are recorded as resident in The Gambia, many are extremely rare or local and are unlikely to be seen by most visitors.

- [ ] Tropical African Hedgehog (Atelerix albiventris)
- [ ] Gambian Fruit Bat (Epomorphorus gambianus)
- [ ] Lesser Galago (Galago senegalensis)
- [ ] Western Baboon (Papio papio)
- [ ] Mona Monkey (Cercopithecus mona mona)
- [ ] Green Vervet Monkey (Callitrix) (Cercopithecus aethiops sabaeus)
- [ ] Patas (Red Monkey) (Erythrocebus patas)
- [ ] Western Red Colobus (Colobus badius)
- [ ] Chimpanzee (Pan troglodytes)
- [ ] Crawshay's Hare (Lepus crawshayi)
- [ ] Gambian Sun-squirrel (Heliosciurus gambianus)
- [ ] Striped Ground Squirrel (Xerus erythropus)
- [ ] Larger Cane Rat (Thryonomys swinderianus)
- [ ] Gambian Giant Rat (Cricetomys gambianus)
- [ ] Brush-tailed Porcupine (Atherurus africanus)
- [ ] Crested Porcupine (Hystrix cristata)
- [ ] Rough-toothed Dolphin (Steno bredanensis)
- [ ] Humpback Dolphin (Sousa teuszii)
- [ ] Bottle-nosed Dolphin (Tursiops truncatus)
- [ ] Common Jackal (Canis aureus)
- [ ] Side-striped Jackal (Canis adustus)
- [ ] Sand Fox (Vulpes pallida)
- [ ] Wild Dog (Lycaon pictus)
- [ ] Zorilla (Ictonyx striatus)
- [ ] Ratel (Honey Badger) (Mellivora capensis)
- [ ] Cape Clawless Otter (Aonyx congica)
- [ ] African Civet (Viverra civetta)
- [ ] Common Genet (Genetta genetta)
- [ ] Large-spotted Genet (Genetta tigrina)
- [ ] White-tailed Mongoose (Ichneumia albicauda)
- [ ] Marsh Mongoose (Atilax paludinosus)
- [ ] Egyptian Mongoose (Ichneumon) (Herpestes ichneumon)
- [ ] Slender Mongoose (Herpestes sanguineus)
- [ ] Banded Mongoose (Mungos mungo)
- [ ] Gambian Mongoose (Mungos gambianus)
- [ ] Spotted Hyaena (Crocuta crocuta)
- [ ] Striped Hyaena (Hyaena hyaena)
- [ ] African Wild Cat (Felis libyca) (may = F. sylvestris)
- [ ] Serval (Felis serval)
- [ ] Caracal (African Lynx) (Felis caracal)
- [ ] Leopard (Panthera pardus)
- [ ] Aardvark (Orycteropus afer)
- [ ] Rock Hyrax or Dassie (Procavia capensis)
- [ ] African Manatee (Trichechus senegalensis)
- [ ] Warthog (Phacochoerus aethiopicus)

☐ Western Bush-pig (Red River Hog) (Potamochoerus porcus)
☐ Hippopotamus (Hippopotamus amphibius)
☐ Sitatunga (Tragelaphus spekei)
☐ Bushbuck (Tragelaphus scriptus)
☐ Bohor Reedbuck (Redunca redunca)
☐ Red-flanked Duiker (Cephalopus rufilatus)
☐ Maxwell's Duiker (Cephalopus monticola maxwelli)
☐ Grimm's Duiker (Sylvicapra grimmia)
☐ Oribi (Ourebia ourebi)

SELECTED BIBLIOGRAPHY

General

Hudgens, J. and R. Trills (1992) West Africa - The Rough Guide. Rough Guides.
Palmer, T. (1991) Discover The Gambia 3rd Edition. Heritage House.
Sweeney, P. (Ed) (1990) Insite Guides - The Gambia and Senegal. APA Publications.

Fauna and Flora

Andersson, L.G. (1937) Reptiles and batrachians collected in The Gambia by Gustav Svensson and Birger Rudebeck. Archiv for Zoologi 29A(16): 1-28
Boorman, J. (1970) West African Butterflies and Moths. Longman.
Cansdale, G.S. (1961) West African Snakes. Longmans.
Carcasson, R.H. (1981) Collins Handguide to the Butterflies of Africa. Collins.
Dorst, J and P. Dandelot (1972) A Field Guide to the Larger Mammals of Africa 2nd Edition. Collins.
Dowsett, R.J. and A.D. Forbes-Watson. (1993) Checklist of Birds of the Afrotopical and Malagasy Regions. Tauraco Press.
Edberg, E. (1982) A Naturalist's Guide to The Gambia. J. G. Sanders.
Gledhill, D. (1972) West African Trees. Longman.
Fry, H., S Keith and E. Urban (Eds.). (1982 onwards) The Birds of Africa. Volumes 1-4. Academic Press.
Gore, M.E.J. (1990) The Birds of The Gambia. B.O.U. Check-list No.3, 2nd (Revised) Edition. British Ornithologists Union.
Hakansson, N.T., (1981) An annotated checklist of reptiles known to occur in The Gambia. Journal of Herpetology 15(2): 155-161
Haltenorth, T. and H. Diller (1977) A Field Guide to the Mammals of Africa including Madagascar (English Translation publ. 1980). Collins.
Jonsson,L. (1992) Birds of Europe with North Africa and the Middle East. Christopher Helm.
Mackworth-Praed, C.W. and C.H.B. Grant (1970) African Handbook of Birds, Series III - Birds of West Central and Western Africa Vol.I. Longman.
Mackworth-Praed, C.W. and C.H.B. Grant (1973) African Handbook of Birds, Series III - Birds of West Central and Western Africa Vol.2. Longman.
Newman, K. (1991) Birds of Southern Africa 3rd Edition. Southern Book Publishers.
Saunders, H.N. (1958) A Handbook of West African Flowers. O.U.P.
Serle, W., Morel G.J. and W. Hartwig (1977) A Field Guide to the Birds of West Africa. Collins.
Sinclair, I., Hockey, P. and W. Tarboton (1993) Illustrated Guide to the Birds of Southern Africa. New Holland.

Wacher, T. (1993) Some new observations of forest birds in The Gambia. Malimbus 15 (1):24-37.
Williams, J.G. and N. Arlott (1980) A Field Guide to the Birds of East Africa. Collins.

LOCAL CONTACTS AND SOCIETIES

Two ornithological societies publish reports relevant to the area. These are:
The Gambia Ornithological Society, P O Box 757, Banjul, The Gambia, West Africa.
The West African Ornithological Society (c/o The Treasurer), 1 Fisher's Heron, East Mills, Fordingbridge, Hampshire, SP6 2JR. Journal: Malimbus.

In addition a new society, the African Bird Club was formed in January 1994 to encourage interest in conservation in Africa. The address for information is:

The Membership Secretary, African Bird Club, c/o Birdlife International, Wellbrook Court, Girton Road, Cambridge, CB3 0NA.

Sightings of rare birds in The Gambia should be reported to Chris Paul, the Recorder of the Gambian Ornithological Society, at the above address, from whom a supply of bird report forms may also be obtained.

Finally, in order to help improve and update this guide, I should be most grateful if any comments, sightings or reports from bird-watching trips to The Gambia could be sent to:

Dr A R Ward, 5 Hawkstone Close, Harwood, Bolton, BL2 3NY, Great Britain.